T0176733

First to File

First to File
Patents for Today's Scientist and Engineer

M. Henry Heines

WILEY

Cover Design: Wiley

Copyright © 2014 by the American Institute of Chemical Engineers, Inc. All rights reserved.

A Joint Publication of the American Institute of Chemical Engineers and John Wiley & Sons, Inc.

Published by John Wiley & Sons, Inc., Hoboken, New Jersey.
Published simultaneously in Canada.

No part of this publication may be reproduced, stored in a retrieval system, or transmitted in any form
or by any means, electronic, mechanical, photocopying, recording, scanning, or otherwise, except as
permitted under Section 107 or 108 of the 1976 United States Copyright Act, without either the prior
written permission of the Publisher, or authorization through payment of the appropriate per-copy fee to
the Copyright Clearance Center, Inc., 222 Rosewood Drive, Danvers, MA 01923, (978) 750-8400, fax
(978) 750-4470, or on the web at www.copyright.com. Requests to the Publisher for permission should
be addressed to the Permissions Department, John Wiley & Sons, Inc., 111 River Street, Hoboken, NJ
07030, (201) 748-6011, fax (201) 748-6008, or online at http://www.wiley.com/go/permission.

Limit of Liability/Disclaimer of Warranty: While the publisher and author have used their best efforts
in preparing this book, they make no representations or warranties with respect to the accuracy
or completeness of the contents of this book and specifically disclaim any implied warranties of
merchantability or fitness for a particular purpose. No warranty may be created or extended by sales
representatives or written sales materials. The advice and strategies contained herein may not be suitable
for your situation. You should consult with a professional where appropriate. Neither the publisher nor
author shall be liable for any loss of profit or any other commercial damages, including but not limited to
special, incidental, consequential, or other damages.

For general information on our other products and services or for technical support, please contact our
Customer Care Department within the United States at (800) 762-2974, outside the United States at
(317) 572-3993 or fax (317) 572-4002.

Wiley also publishes its books in a variety of electronic formats. Some content that appears in print may
not be available in electronic formats. For more information about Wiley products, visit our web site at
www.wiley.com.

Library of Congress Cataloging-in-Publication Data:

Heines, M. Henry, 1945– author.
  First to file : patents for today's scientist and engineer / M. Henry Heines.
    pages   cm
  "A Joint Publication of the American Institute of Chemical Engineers and John Wiley & Sons, Inc."
  Includes index.
  ISBN 978-1-118-83965-2 (cloth)
1. Patent laws and legislation–United States.   2. Patent practice–United States.   I. Title.
  KF3114 .H447
  346.7304'86–dc23

                                                                                    2014012664

Printed in the United States of America.

10   9   8   7   6   5   4   3   2   1

Contents

List of Figures

List of Tables

Preface

What is the scientist's or engineer's role in the patenting process? Why not leave it all to a patent attorney? On the other hand, why use a patent attorney at all?

As this book will demonstrate, patent law has its underpinnings in technology. What the patent statute lists as the qualifications for a patentable invention, augmented by the definitions and interpretations imposed by the courts that hear patent cases, all reflect an interest in recognizing and rewarding technological advances and in discriminating those advances that are meritorious from those that are mere trivialities. Indeed, both the patent statute and the court decisions make repeated reference to the "person having ordinary skill in the art" as the ultimate arbiter of patentability. The technologist's view as to what this hypothetical person might find "obvious" or "nonobvious" adds a measure of realism to this highly debated principle of patent law. Clearly, the technologist's participation in the patenting process is both essential and advantageous.

The patenting process is also of interest to managers, officers and directors, and anyone in a technology-based organization whose function is to promote the viability, sustainability, and the growth of the organization. As some of the chapters in this book will demonstrate, numerous statutory provisions and regulations relating to patents reflect general business practices, such as the rights of employers versus employees and vice versa, and the economic realities of running a corporation and competing in the marketplace. The diligent patent attorney will make the appropriate inquiries in the interest of providing the corporation with defensible patent protection, but the corporate officer can avoid a potential loss of rights by anticipating the questions.

Who actually makes patent law? The patent statute provides the rudiments, but the definitions and interpretations as they apply to different areas of technology are made by courts of law in the decisions rendered in patent lawsuits. The federal judges rendering these decisions are generally not scientists or engineers, and although many of them have a working familiarity with technological matters, they generally lack the benefit of an up-to-date knowledge of the latest technological advances. To perform their jobs, judges rely on attorneys and expert witnesses for information as to what the technology community considers to be the state of the art. The attorneys, however, have their clients' interests to protect, and even expert testimony will contain some degree of bias. As a result, the arguments and explanations that the judges hear are neither entirely nor consistently objective. It is the technologists themselves who are most likely to be true to the underlying technology and are therefore in the best position to respond to technological questions. A technologist who also understands the legal issues and standards set by the courts will be able to address the technological issues directly and to provide the most relevant advice and insight.

The informed inventor will select a patent attorney with the relevant technical education so that inventor and attorney can communicate at a high level about the invention, and so that the attorney can present the invention to its best advantage to the Patent and Trademark Office (PTO) and the courts. Each invention is an advance in the state of the art, however, and the attorney needs the inventor's insight in defining the state of the art and identifying the need for and significance of the invention, in addition to explaining the details of the invention itself. Here again, it is the inventor/technologist who is in the best position to do this, and can do it best with an understanding of the issues that the attorney will face in applying for a patent.

The most intriguing (and sometimes challenging) inventions are those in emerging technologies, since these technologies by their nature are either unexplored or in the early stages of their exploration. Fitting them into entrenched legal frameworks so that they can achieve the kind of legal protection that they deserve can only be done with the technologist's help. Many emerging technologies require the development of new legal theories and new ways of applying established legal theories. This can only be done with an intimate understanding of the new technologies, and of their implementations and potentials, all of which are within the province of the technologist.

Even if your attorney has extensive experience in patent law and a detailed technical knowledge in the field of your invention, should you place all your trust in the attorney and allow your attorney to make all decisions on your behalf? Would you give your investment advisor or personal asset manager complete control over the management of your assets without consideration of your personal interests, concerns, or preferences? Would you allow your real estate agent to select a house for you and your family or a neighborhood for you to move into without your input? Since patents are about technology, the inventor/technologist will in most cases have a more detailed understanding of the subject matter than the attorney and a personal interest in the type and scope of coverage to be obtained. The technologist may also be called upon to explain or defend the claims and arguments that have been made before the PTO or a judicial authority. There is no need or justification for blindly entrusting everything to the attorney, no matter how competent.

In my 40 years' experience as a patent attorney, I have found scientists and engineers to be very intelligent people, and inquisitive as well. Individuals of that caliber are fully capable of understanding the fundamentals of patent law at a high level. The "Dick-and-Jane" approach that some writers use in teaching potential inventors about patents is inappropriate, if not demeaning, to these individuals, particularly those with advanced degrees in technical disciplines and those with years of hands-on experience in technical fields, and the same is true for management professionals in technology companies. This book does not attempt to cover all topics in patent law, but for those that are covered I have endeavored to answer the next logical question in each case and to give a reasonably complete expression of what the legal parameters are and where they came from.

Finally, if the technologist's involvement is so important and the technologist is so capable intellectually, why involve a patent attorney at all? (I use the term "patent attorney" for convenience, but I intend it to include patent agents, even though this is not strictly the correct usage. Patent attorneys and patent agents are both licensed to represent inventors before the PTO, but a patent attorney is also licensed to provide patent analyses and opinions as well as other legal services beyond representation before the PTO.) Why shouldn't the technologist serve as her/his own patent attorney and deal directly with the PTO by communicating directly with the PTO administrative staff and the examiner and preparing any necessary petitions and administrative appeals? Books have certainly been published with the aim of enabling inventors to act as their own patent attorneys, and certain inventors have done so successfully.

Nevertheless, the representation of inventors before the PTO is a practice of law, and in fact a specialty within a specialty, requiring a certification from the PTO aside from any certification from a state bar to practice law in general. For this reason, general practice attorneys are typically reluctant to provide patent advice, and most often refer their clients to a qualified patent attorney when patent issues come up. Like any field of law, the practice of patent law includes legal analysis and argument, which is more than selecting appropriate forms and filling them out properly or preparing documents with regulation-prescribed formats. The legal skills that a qualified patent attorney can offer include the ability to present your best case to the PTO and do so with a long-term perspective, that is, in a way that does not compromise your coverage years later when shortcomings and other flaws that are no longer correctable are revealed. A patent attorney can provide current perspectives on legal developments and on any changes in legal standards that are adopted by the courts and the PTO, as well as the experience of dealing with examiners and the PTO in a multitude of cases over a span of time. As for the forms and formats, these change, and patent attorneys are in the best position to know when changes have been made as well as which forms to use for any contingency. Self representation is often proposed for individuals who have a "gizmo" or a "nifty idea" that they believe to have potential as a consumer item that will catch on quickly when revealed and yet do not wish to make a large investment in securing legal protection. The most valuable inventions however are not gizmos or nifty ideas, but instead relatively sophisticated innovations in fields that are well-populated with new ideas and improvements that continually advance the state of the art, and the most prolific patent filers, corporate or otherwise, use professional patent attorneys, either in-house or independent contractors. Regardless of the nature of the invention or the field to which it applies, the chances of becoming a millionaire on a de minimis investment in legal services are de minimis themselves.

Nevertheless, the PTO regulations do allow for self (*pro se*) representation and many individuals do succeed on their own. This book is not however intended as a resource for the self-representing inventor, but rather to enable the inventor to get the most out of the patent system and the attorney–client relationship and to thereby obtain the most durable and valuable patent protection, particularly under the America Invents Act of 2011.

Introduction

The number of books currently available on the subject of patents numbers in the many tens of thousands, with authors drawn from the ranks of attorneys, academics, officials of the U.S. Patent and Trademark Office, and even inventors. The approaches vary widely, ranging from basic introductions to patents and intellectual property in general to guides for attorneys, inventors, and business professionals. How-to books for nonprofessionals, books on searching, books on the valuation and marketing of patents, and much more abound. Why then another one?

One reason is that the patent system is in continuous flux. New technologies are continually testing time-honored principles of patent law and challenging the reasoning behind them; new forms of media and information exchange are changing the definition of novelty and the state of the art; evolving interests of business and commerce are challenging long-held notions of what should and should not be allowed as patentable subject matter; an increasing need for international cooperation in defending and enforcing intellectual property leads to a gradual elimination of the differences in intellectual property laws among different countries; new precedents are set as judicial attitudes change and evolve; and questionable practices in the use of patents (such as the controversy over "trolls") take on a new prominence. Any book on patent law is therefore to some extent a snapshot, and periodic revisiting and updating are essential to maintaining a working knowledge of the subject. This is even more critical now with the enactment of the Smith-Leahy America Invents Act of 2011 (AIA) and its institution of some of the most fundamental changes in patent law and procedures since the patent system was first created.

The most widely debated and controversial change made by the AIA is the switch from first-to-invent to first-to-file as the rule governing the choice between competing inventors applying for patents on the same invention. I begin this book with a description of the new rule and how it applies to priority disputes, and then proceed with the changes in the definition of prior art as reflected by the rule as well as the changes to prior art under the AIA that are not a direct consequence of the first-to-file rule. These are followed by explanations of the business-related features of the AIA, such as the adaptation of patent procedures to more realistically reflect the business environment and the new options introduced by the AIA for challenging patents of questionable validity. Interspersed within these topics are outlines and illustrations of the fundamentals of patent law as they apply to the scientist or engineer (i.e., the technologist in general), using actual cases to demonstrate the fundamentals and to shape the technologist's perspectives on and approaches to intellectual property, particularly those perspectives that have a bearing on the technologist's activities before the first contact with patent attorney or before starting the patenting process.

Particular emphasis is placed on recent court cases showing the current attitudes of the Patent and Trademark Office and the judiciary on the various standards of patent law. Included among these are areas that are matters of active debate and likely to evolve further in the years to come, such as the patentability of business methods, computer-implemented inventions, and medical diagnostics involving the functions and uses of human genes.

Finally, I include a chapter that places the fundamentals of patents and patenting in the context of intellectual property in general, including those other intangible assets that are not strictly alternatives but rather complementary to patents and are part of the valuation and asset portfolio of the successful and competitive business or entrepreneur. Trade secrets, trademarks, copyrights, and patents other than utility patents offer their own unique forms of protection and value, and comparisons among them will help strengthen one's understanding of the patent system and intellectual property as a whole.

About the Author

M. Henry Heines is a patent attorney, consultant, and author and speaker on patent-related topics, with a publishing history that includes over 60 articles in technical and legal journals and two books on patent law for the technical business community. The present book is his third, complementing and expanding on the first two to address the recently enacted America Invents Act, and to include other topics of current and emerging interest, supplemented with actual case histories, that show the current state of the U.S. law of patents and the evolving directions of the law as indicated by current legal developments. In addition to his legal credentials, the author holds a Ph.D. in chemical engineering supplemented by additional course-work in molecular biology and immunology, and has practiced patent law in a wide range of fields including chemistry, chemical and mechanical engineering, labora-tory, industrial, and medical equipment, materials science including metallurgy and nanotechnology, and alternative fuels and energy sources. Recently retired from the full-time practice of law, his professional career began with three years as a research engineer with a major chemical corporation, followed by seven years in corporate patent practice, and finally 31 years in private practice in the firm now bearing the name Kilpatrick Townsend & Stockton LLP. Dr. Heines currently resides in San Francisco, California, and maintains a personal website www.henryheines.com.

Chapter 1

The First-to-File Rule: Evolution and Application

Senator Patrick Leahy and Representative Lamar Smith, the Congressional sponsors of the America Invents Act (**AIA**), when considering the volume and intensity of the debate that preceded the signing of the Act into law in 2011, may have found it amusing, if it occurred to them at all, that the first U.S. patent examining body included among its three members the Secretary of War, Henry Knox.[1] While Mr. Knox, who has since come to be recognized as the father of American army artillery, possessed considerable war expertise, it is doubtful that this expertise was a factor in his being appointed to the Patent Board, and it is safe to say that **patents** have never been at the center of a military war. Nevertheless, patents have generated controversies at various times since their introduction into the United States, and possibly the most hotly debated controversy was the one over the proposal made 200 years after Mr. Knox assumed his duties as an examiner, that is, that patents be awarded on a "first-to-file" basis rather than the long-standing policy of "first-to-invent." Those opposing the proposal argued that, among other perceived evils, it would reverse 200 years of precedent, and vigorous arguments both for and against the proposal were expressed not only by various interest groups within the United States but also between the United States and its allies. The proposal passed however, and it, together with the other provisions of the AIA, produced the greatest overall change in U.S. patent law in 60 years.[2]

The "**first-to-invent**" and now the "**first-to-file**" **rules** were devised to resolve priority disputes, that is, competing attempts to obtain patent coverage by different individuals or entities who have separately invented the same invention. Although

---

[1] The other two members of the Patent Board (or, as it called itself, the "Commissioners for the Promotion of Useful Arts") were Thomas Jefferson, who was the Secretary of State, and Edmund Randolph, who was the Attorney General.

[2] The most recent major change occurred with the enactment of the U.S. Patent Act of 1952.

First to File: Patents for Today's Scientist and Engineer, First Edition. M. Henry Heines.
© 2014 the American Institute of Chemical Engineers, Inc. Published 2014 by John Wiley & Sons, Inc.

these priority disputes could seemingly arise anywhere in the world and the U.S. patent system was not the world's first, the United States had to develop its own rule with little guidance from preexisting systems.

1.1   HISTORY OF THE FIRST-TO-FILE RULE IN THE UNITED STATES

The patent system existing in England at the time that the empowering clause in the U.S. Constitution[3] was written and the first U.S. patent law[4] was enacted is commonly considered the basis for the U.S. law. The original English patents were privileges granted by the Crown under royal prerogative rather than property rights and were not rewards for ingenuity or discovery. In fact, the first patents were patents of importation, granted to individuals to reward them for introducing products and processes into the country from abroad. This was soon expanded to include patents of invention, that is, for innovations originating within the country itself, but it eventually became apparent that both these types of royal grants were more of a hindrance to domestic industry at large than an incentive for technological advance. As a result, Parliament enacted the Statute of Monopolies in 1621, which voided all patents, including patents of invention and importation as well as business licenses, *except* those for the "sole working or making of any manner of new manufacture within this realm, [granted] to the first and true inventor or inventors … [that was] not contrary to the law nor mischievous to the state by raising prices of commodities at home, or hurt of trade …." While the wording of this exception would appear to be an explicit limitation to patents of invention and an institution of a first-to-invent policy, these were both subsequently obscured by the English courts in their interpretation of the expression "the first and true inventor or inventors" to include importers of products and processes that the importers had not themselves invented. Whatever the expression may have implied, however, no priority disputes were adjudicated in England between Parliament's enactment of the Statute of Monopolies and the U.S. Congress' enactment of its first Patent Act almost 170 years later. This left the United States with no precedent on how to resolve priority disputes other than the Statute of Monopolies itself, which was obscured by the legal system's loose interpretation of "the first and true inventor."

Unlike England, the United States in the late eighteenth century was forced to confront the issue since it faced a situation not present in England. Whereas patents in England were granted by a central authority, that is, the Crown or its law officers, the American colonies lacked a central authority and were granting patents individually well before the Revolutionary War. Even though the colonies claimed to have received the authority to do so from their colonial charters, the colonies were not consistent in how they interpreted their patenting authority. Some colonies granted patents of invention and not importation, some granted patents of

---

[3] Article 1, Section 8, Clause 8, enacted in 1787.
[4] The Patent Act of 1790.

importation and not invention, some granted both, and some refused to grant any patents. The inconsistencies remained as the colonies became states, but were partially mitigated by the short-lived Articles of Confederation (1781). The Articles continued to recognize the power of individual states to grant patents, however, and defined infringement to include acts occurring within the granting state as well as the importation of infringing products from other states. As for priority among competing inventors, an inventor could clearly be the first to file in one state and the second in another, and yet an early filing in one state by a particular inventor could serve as evidence of that inventor's early invention. It soon became apparent that challenges to patent validity and enforcement among different states were awkward to reconcile and that patents were of no practical use unless they were equally enforceable in all states. Furthermore, the growth in interstate commerce and the need to develop domestic industry in competition with the importation of foreign goods soon took precedence over any interest in individual state patents. For these reasons, the value of rewarding the patent to the first to invent rather than the first to file was apparent.

Countries outside the United States likewise developed their patent systems individually, based on their own interests. Although the twentieth century saw the enactment of patent treaties between groups of countries for various reasons, the initial creation of most worldwide patent systems occurred without efforts of individual countries to band together. Economic competition between countries may in fact have caused individual countries to place a high value on early filing, both to introduce new technology into their countries at an early date by way of the descriptions in patents and to obtain early expiration dates for their patents to hasten the release of the new technologies to the public. This latter goal was achieved by setting the expiration date of a patent at a fixed number of years from the filing date, a policy that the United States did not adopt until 1995.

First-to-file thus became the general rule worldwide, with the United States being the sole exception. Why then did the United States hold out for so long? There is certainly no reason to expect that two (or more) individuals, with or without knowledge of each other's existence, were any more likely to come up with the same invention in the late twentieth century than they were in the eighteenth. Individuals in the late twentieth century were more likely to seek patent coverage, however, due to their recognition of the greatly increased economic power of patents, and the administrative complexities of determining the first to invent in priority disputes became ever more cumbersome and expensive both for the parties involved and for the Patent and Trademark Office (**PTO**). Most significantly, however, growth in the global economy and international trade placed the United States under pressure to reconsider its adherence to the **first-to-invent rule**, since the adverse effects of this growth included an increasing trade deficit in the United States as well as the piracy of American products by manufacturers in third world countries. The resulting damage to U.S. companies and the U.S. economy in general has prompted the United States to try to enforce its intellectual property more aggressively, as evidenced by an increase in U.S. inventors applying for patents abroad and a desire for U.S. patents to have a more global impact. International treaties are an effective means of promoting these goals, by harmonizing standards of patentability, coordinating procedures for applying for patent protection in multiple countries, and simplifying the means of enforcing patents across international boundaries. Both Congress and American industry have recognized the potential benefits of these goals, and the United States has entered into such treaties, including the **General Agreement on Tariffs and Trade**,[5] the **Trade-Related Aspects of Intellectual Property**,[6] the **World Trade Organization**,[7] and the **World Intellectual Property Organization**.[8] Nevertheless, full participation of the United States in seeking global harmonization has been hindered by the U.S. adherence to the first-to-invent rule. The AIA and its institution of the first-to-file rule thus remove this obstacle.

The controversy that preceded enactment of the change by the AIA reflects the profundity of the change relative to other features of U.S. patent law, as evidenced not least by the complex set of official regulations and procedures implementing the first-to-file rule and the fact that the expertise needed to navigate these regulations has created its own subspecialty among patent attorneys. One argument against the first-to-file rule was that it is unconstitutional, based on an interpretation of use of the word "inventors" in the empowering clause[9] of the Constitution to mean "true inventors" and therefore "first inventors." Another is that the rule unfairly benefits large, well-funded corporations over individual inventors, startups, and nonprofit entities such as universities and research institutions, the large corporations presumably being better able to fund multiple and early patent filings. The constitutionality argument has been less than fully compelling, however, since the later of two inventors can still be a "true" inventor, and the Constitution does not state otherwise, and the empowering clause expressly states that the purpose for empowering Congress to enact patent laws is "[t]o promote the Progress of Science and useful Arts" and therefore to provide an incentive for inventors to promptly file their patent applications so that the public can obtain the greatest benefit from the information contained in the patents. As for the perceived unfairness to individuals, small businesses, universities, and research institutions, this is partially mitigated by the fee discounts that the PTO offers for "**small entities**" and "**micro entities**," as well as administrative procedures introduced by the AIA that provide faster, cheaper, and more streamlined means of challenging patents and adjudicating patent disputes, all of which can benefit applicants who are less well funded. Ultimately, the fact that the United States has been the sole holdout among patent-granting countries of the world by adhering to the first-to-invent system, combined with the United States' global economic considerations, is most responsible for instituting the change.

---

[5] **GATT**, 1948.

[6] **TRIPS**, 1994.

[7] **WTO**, 1995.

[8] **WIPO**, 1967 (United States joined in 1970).

[9] Article 1, Section 8, Clause 8: "To promote the Progress of Science and useful Arts, by securing for limited times to Authors and Inventors the exclusive Right to their respective Writings and Discoveries...."

1.2 "WHO'S ON FIRST?": THE RULE AND ITS APPLICATION

The first-to-file rule applies to competing *inventors* who file patent applications, not simply to competing *filers* (i.e., applicants). Any individual can file a patent application on another's invention by substituting the filer's name for that of the true inventor, whether for unscrupulous reasons or because of a lack of understanding of the law. The law requires that the true inventor be named when submitting the application, and a failure to do so, for whatever reason, is grounds for an examiner to reject an application or for a court to declare a patent invalid. Competing *inventors*, however, do not risk rejection or invalidation simply because of the competition, provided that each is a true inventor. They do however confront each other in a priority dispute that is resolved in favor of the first among them to file.

The act of invention, under either the first-to-invent rule or the first-to-file rule, is generally the **conception** of an idea followed by the reduction of the idea to practice. For certain inventions, the **reduction to practice** is a routine matter involving no creative input or other contribution to the conception once the conception is made, and some inventions lack an actual reduction to practice and instead are sufficiently well thought out to allow the direct filing of a patent application without construction of a prototype or the generation of test data. For these inventions, the act of invention is the conception itself. In either case, many individuals at dispersed locations obtain their education and expertise in the same area of technology, work in the same industry, read the same technical publications, and recognize the same problems in need of solution. Occasions will therefore arise where two or more individuals independently conceive of, and act upon, the same inventive idea. When independent inventors or independent groups of inventors apply separately for patents under the first-to-file rule, the rule applies regardless of whether either one was aware of the other's existence, or of the fact that the other was working in the same area of technology, or even of the fact that the other intended to apply for a patent. If any such awareness is present however and it can be shown that a particular applicant filed an application on an invention obtained from, or derived from information obtained from, someone else, recourse for the one who originated the idea is available through a "**derivation proceeding**" rather than a simple application of the first-to-file rule. A "derivation proceeding" is an administrative proceeding conducted within the PTO (or **USPTO**) and was introduced by the AIA. The requirements and procedures for derivation proceedings are addressed in Chapter 4 of this book.

The term "file" in its various grammatical forms appears throughout both the patent law and the PTO regulations that implement the law, both before and after the AIA. The inventors will therefore benefit from knowing what constitutes a "filing" under both the first-to-invent and the first-to-file rules. The term "filing" generally refers to a submission of a patent application to an agency created by a government or by an international treaty to receive the application. The agency will assign the application an application number that is unique to the application and will issue the submitter a receipt indicating the application number and the official date of receipt. The USPTO is one example of such an agency, patent offices (under various names) in individual countries outside the United States are further examples, and patent offices established by treaties among certain groups of countries to serve as centralized patent agencies, prominent among which are the European Patent Office (**EPO**) and the **Patent Cooperation Treaty** (**PCT**), are still further examples. The treaty with the largest number of countries is the PCT, according to which applications can be filed in any of several receiving offices authorized by the treaty (of which the USPTO is one). The term "filing" does not include the submission of documentation to a patent attorney, to a patent administrator (such as a staff person in a corporate patent or legal department), or to any other governmental authority (such as the occasionally but misguidedly cited mailing of a description of the invention to oneself to obtain a receipt stamp from the U.S. Postal Service). Of the agencies which "filing" does include, the functions performed by these agencies differ; some are empowered to grant patents, while others simply receive and record the applications and perform a preliminary processing that is later continued by the same agency or by another agency toward granting a patent. The "applications" that can receive "filing" status include both those that will ultimately become patents and those that serve as predecessors to applications that will become patents. These predecessor applications include **provisional patent applications** (defined and described in Chapter 10) that are filed in the USPTO but not examined, PCT applications, patent applications filed in jurisdictions outside the United States and then followed with U.S. counterparts, and nonprovisional U.S. patent applications that are refiled in the USPTO for purposes such as expansion, updating, changes in emphasis, and the opportunity to present more argument and renew the examination process.

"Filing" thus covers a variety of documents submitted to a variety of authorized receiving offices, and under appropriate conditions (including timing and documentation), two or more such filings on the same or a closely related invention by the same inventor can be made in succession in the USPTO. When such successive filings are made, the later filing(s), typically the refiled applications mentioned earlier, will commonly reference the earlier one(s) with all of the applications claiming the benefit of the filing date of the earliest-filed application of the series. The terms "**continuation**," "**continuation-in-part** (**C-I-P**)," "**divisional**," and "**reissue**" are applied to many of these refiled applications. These types of refiling are common practice, and the strategies behind their use are all within the expertise of a patent attorney. In each case, however, the earliest claimed filing date is then the "**effective**" **filing date**, while the actual receipt date of any application in the USPTO is that application's "**actual**" **filing date**. The legally sanctioned use of the benefit of an "effective" filing date that precedes the actual filing date was well established prior to the enactment of the AIA but has been expanded significantly under the AIA and given increased emphasis by the AIA's explicit use of the expressions "effectively filed" and "effective filing date" in the statute itself. The "first-to-file" thus means the first to *effectively* file.

1.3 ADAPTING BUSINESS ROUTINES TO THE FIRST-TO-FILE RULE

Any technology-based company or organization that seeks financial stability and growth will have an intellectual property policy that includes record keeping, control of outside disclosures and maintenance of confidentiality, and strategies for securing intellectual property rights with an emphasis on patents. Should any of these practices be expanded, eliminated, or changed in view of the first-to-file rule introduced by the AIA?

The date on which the first-to-file rule came into effect was March 16, 2013. The rule is not retroactive and therefore does not apply to applications pending on that date, to applications filed after that date but with effective filing dates before that date, or to patents granted on applications with effective filing dates before that date. All such applications and patents are still governed by the first-to-invent rule, and the rule will therefore continue to have an impact until the last patent subject to the rule expires or is no longer in force. This impact is shown in Chapter 2. For these first-to-invent cases, the value of diligent record keeping is unchanged by the AIA. Record keeping does lose a certain degree of significance in first-to-file cases, but the loss is only in the area of priority disputes. For reasons aside from priority disputes, such as establishing collaboration with others outside the company and either joint or outright ownership (see Chapter 4), as well as derivation (also addressed in Chapter 4), record keeping not only retains its value under the new rule but in some cases has greater value.

For purposes of priority disputes, early filing clearly has greater value under the first-to-file rule. Does this make it more difficult for individuals and small corporations to secure patent protection, as argued prior to the AIA's enactment? The argument stresses that individuals and small corporations are in a weaker position to manage the costs of early filings, particularly for inventions that are at a rudimentary stage and not sufficiently tested to determine technological or economic viability, and that individuals and small corporations have fewer resources to accelerate the development of an invention and to assess its economic value before filing. It has also been argued however that smaller companies have fewer inventors and therefore fewer inventions, with a correspondingly lower frequency of patent application filings. This also means less of a need for the services of patent attorneys and hence fewer legal bills. Likewise for individuals and small groups, different inventions are likely to be related in subject matter, simplifying the process of preparing the applications. This is in addition to the "**small entity**" and "**micro entity**" discounts mentioned earlier.

Nevertheless, the business routine can be adapted to protect against or minimize any loss of opportunities under the first-to-file rule. Company policy can be adapted by raising patent filings to a level of priority comparable to that of record keeping, and this can be done by including a patent attorney, or a staff person assigned to coordinate or administer patent matters with outside patent counsel, in the distribution lists of internal company research reports and in presentations or meetings where research results are discussed and evaluated. To ensure that ongoing developments are reported, inventors should keep copies of their patent applications and consult them periodically to compare their latest research efforts and developments to those described and covered in the existing applications and to see where updating is needed. Inventors should also be encouraged to think broadly when discussing their inventions with patent counsel. An early filing date is of greatest value when the scope of the application as filed extends beyond the immediate area of the invention's interest to the inventor or the company. This can be done by first identifying the central distinguishing feature of the invention and expanding its scope of possible implementation beyond that which prompted the invention, even to areas that are well removed from the actual laboratory work and even if the chances of viability in those areas are speculative. Chapter 10 expands on this.

Among academic researchers, the publishing of one's research is often an integral part of building one's career and professional reputation, and in many cases, publishing also contributes to one's chances for long-term academic employment. The implementation of the first-to-file rule raises the question of whether the potential for patent protection for academic researchers or the institutions that employ them must suffer if the institutions continue to allow their members to publish. The answer is no; the 1-year grace period allowed to inventors for publishing their inventions that has long been a feature of U.S. patent law remains in effect under the AIA and in fact is expanded to include publications by others who have obtained their subject matter directly or indirectly from the inventor. This is explained more fully in Chapter 4. Disclosures and publications outside the grace period, that is, more than 1 year before the effective filing date of the patent application, are as much of an obstacle to patentability before the enactment of the AIA as after. And as mentioned earlier, derivation proceedings are available to inventors whose inventions have been misappropriated by another who files an application under the filer's own name but based on the inventor's disclosure or publication.

Publication of an invention before filing will nevertheless restrict one's ability to patent the invention outside the United States, that is, in countries that do not provide for grace periods, and this is true under both first-to-invent and first-to-file rules. Those companies or institutions that have instituted policies that control their members' permission to publish or that coordinate their members' publishing activities with patent filing procedures in a manner designed to preserve patent rights will thus continue to benefit from these policies. Indeed, they will benefit more under the first-to-file rule, since early filings will avoid the need for or reduce the chances of having to endure the cost and burden of derivation proceedings or of any other efforts to establish that the author of the publication obtained the published information from the inventor.

Chapter 2

Prior Art before and after the AIA: Two Standards Compared

Priority disputes, the subject matter of Chapter 1, are conflicts between patents or patent applications (or between a patent and a patent application) in a common jurisdiction (such as the United States) that claim the same invention, and the resolution of a priority dispute is a determination of which party will be granted the patent (or which party's patent will be enforceable). A subject separate and distinct from priority is patentability, which focuses on the invention itself rather than when, in what form, or by whom a patent application is filed. To be patentable, an invention must be novel, useful, and nonobvious. Of these three requirements, usefulness (**utility**) is easily met and rarely a matter of dispute, but **novelty** and **nonobviousness** have acquired complex definitions embodied in extensive sets of rules and parameters that have evolved with the advancement of both technology and communications first in the Industrial Age and then in the Information Age. Novelty and nonobviousness are the subjects of Chapters 6 and 7 of this book, respectively, and both refer to comparisons between an invention and the "**prior art**."

2.1 PRIOR ART AND THE FIRST-TO-FILE RULE

"Prior art" is defined as everything that an invention can be compared to when determining whether the invention is worthy of a patent. Prior art thus goes beyond patents and patent applications that raise priority disputes to include a large body of knowledge and activity in a wide variety of forms. The scope of this body is set forth by statute rather than, for example, economic factors or moral standards. Included within the statutory scope are subject matter that is both published and unpublished and subject matter that is widely known as well as that known only to a small circle of individuals. Information that is highly developed and of proven accuracy, viability, or

First to File: Patents for Today's Scientist and Engineer, First Edition. M. Henry Heines.
© 2014 the American Institute of Chemical Engineers, Inc. Published 2014 by John Wiley & Sons, Inc.

reliability is included, but prior art can also extend to information that has not been tested or verified. And while knowledge that has been known for years, decades, or more will be prior art, knowledge known only for a matter of days may also be prior art.

The patent statute divides prior art into categories, which include other patents and patent applications that claim the same invention, other patents and patent applications that disclose the same invention without claiming it, published literature and public disclosures revealing the invention, and commercial activity involving use or other exploitation of the invention. Prior art also extends beyond subject matter that is the same as the invention to include any knowledge or disclosure that would indicate or suggest that the innovation embodied in the invention is less than one that would "promote the Progress of Science and useful Arts" to a degree that would justify the grant of a patent. Thus, while an invention may be an acknowledged departure from all that preceded it, the preceding matter is still prior art and the departure is evaluated for its magnitude, significance, import, or effect before a patent will be granted on the invention. Otherwise stated, prior knowledge or activity that poses questions of nonobviousness as well as prior knowledge or activity that poses questions of novelty can both be prior art.

While the first-to-file rule of the AIA is best known as the means by which priority disputes that fall under the AIA are resolved, the rule extends further by modifying the definition of prior art as a whole. Patent applications filed on or after March 16, 2013, will therefore face prior art defined by the first-to-file rule, while the prior art for those filed before that date will still be defined by the first-to-invent rule. Certain applications filed after March 16, 2013, will still be subject to the first-to-invent rule, however, and some will be subject to both. This chapter sets forth the differences between the two definitions of prior art and shows how to determine which one will govern for a given application. First, however, it is helpful to know the categories of prior art, that is, which types of materials and activities constitute prior art, under either the first-to-invent rule, the first-to-file rule, or both.

2.2   "BUT IS IT ART?": THE *ART* OF PRIOR ART

The prior art that is most often encountered and most readily recognized and defined, and in many cases most readily located, are U.S. patents and published U.S. patent applications. These patents and patent applications are prior art regardless of what they claim as their own inventions. It is a common occurrence for a description or other information that is not covered by the claims of a patent (or application) to nevertheless appear in the body of the patent and to have a bearing on the novelty or nonobviousness of what a different individual may later claim as an invention. The information may, for example, appear in a background section of the patent or in a comparative demonstration included among the patent's working examples, with the patentee either failing to recognize or acknowledge the novelty, potential patentability, or potential value of the information. The information may also have been something that the patentee had initially included in the claim scope but later reworded the claims to exclude. The reason for trimming the claims in this manner may have been a loss of confidence in the invention's utility when expressed in its original broad scope or a decision to narrow the scope of the claims in order to expedite the claims' approval and hence the issuance of the patent.

A second category of prior art is published literature such as news media, technical journals, technical data sheets and other product literature, books and electronically published materials in general, archived reports, treatises, and academic theses, and any materials or information that have been made available to the public at large or to those in the relevant field of technology without confidentiality restrictions. This includes patents granted outside the United States and patent applications published by bodies other than the United States Patent and Trademark Office (USPTO). A third major category is commercial activity, which includes use in manufacturing, sales, and offers for sale.

A fourth category, introduced by the AIA, is disclosure that is "otherwise available to the public." Other than by these four words, this category is not defined by the AIA but is characterized by the Patent and Trademark Office as a "catch-all provision" that "permits decision makers to focus on whether the disclosure was 'available to the public,' rather than the means by which the claimed invention became available to the public."[1] Examples listed by the Patent and Trademark Office include a student thesis in a university library, a poster display or other information disseminated at a scientific meeting, subject matter in a laid-open (i.e., published) patent application, a document electronically posted in the Internet, and a commercial transaction that does not constitute a sale under the Uniform Commercial Code. The "otherwise available to the public" category and the published literature category thus overlap.

2.3 AND IS IT "PRIOR"?: PRE-AIA LAW VS. THE AIA

The "art" of prior art thus differs between pre-AIA law and the AIA, but the AIA is best known for its introduction of the first-to-file rule, which imposes its greatest differences on the "prior" of "prior art," with more art being "prior" under the AIA than pre-AIA. Both standards have their exceptions, however, and the topic is best addressed by examining each category of prior art individually.

2.3.1 U.S. Patents and Published U.S. Patent Applications

For clarity, the patents and patent applications that serve as prior art will be referred to herein as "reference" patents and patent applications to distinguish them from patent applications that they are cited against. These reference patents and patent applications are those that name another individual as inventor, that is, someone other than the inventor seeking to overcome them as prior art, and this is true regardless of whether in addition to the "other" inventor the reference also names the inventor against whose application they are cited. As prior art, reference patents and

---

[1] "Examination Guidelines for Implementing the First Inventor to File Provisions of the Leahy-Smith America Invents Act," published by the United States Patent and Trademark Office on February 11, 2013. Although the statement seems to distinguish between "the disclosure" and "the claimed invention," it would appear that both expressions refer to any information whose subject matter raises questions regarding the novelty and nonobviousness of the invention.

reference published[2] patent applications are treated the same. Both achieve their prior art status upon *publication*,[3] but once these documents have achieved that status, their dates as prior art references are their *effective filing dates*.[4] Under pre-AIA law, the inventor whose application is rejected over one of these references could remove the reference from consideration as prior art regardless of its content, and would therefore no longer need to distinguish over it, by showing that the inventor had an earlier date of invention. Such a showing most often consists of documentation of the inventor's own conception of the invention at a date preceding the reference's effective filing date, provided that no more than 1 year had passed between the publication of the reference and the inventor's effective filing date.

Under the AIA, the inventor can remove the reference patent or application from consideration as prior art only by showing that the inventor's *own effective filing date* precedes the effective filing date of the reference. To avoid having to show that her/his invention bears a patentable distinction (i.e., is both novel and nonobvious) over the reference, therefore, the inventor will have to have filed sooner under the AIA than under pre-AIA law; the inventor can no longer go back to the inventor's conception date to predate the reference. There are three exceptions to this rule, that is, three scenarios in which despite the reference's earlier effective filing date, the inventor can still remove the reference from consideration as prior art. One is by showing that the subject matter of the reference was obtained directly or indirectly from the inventor. The second is by showing that the inventor himself or herself, or one obtaining the subject matter directly or indirectly from the inventor, had publicly disclosed the subject matter first. The third is by showing that the subject matter of the reference and that of the inventor's patent application were owned by the same person (or entity) or that the inventor named in the reference and the inventor against whom the reference is cited were both under an obligation (usually by contract) to assign their inventions to the same person or entity. There is however an exception within each of these exceptions: if the publication of the reference (i.e., the issue date if the reference is a patent and the publication date if the reference is a published application) occurred more than 1 year before the effective filing date of the application that the reference is cited against, the earlier filing date of the reference cannot be overcome regardless of whether the subject matter of the earlier-filed reference was

---

[2] Publication in this context means publication by the USPTO.

[3] The issuance of a patent is its publication, while an application is published approximately 18 months from its earliest effective filing date regardless of whether it ultimately becomes a patent. With rare exceptions, all nonprovisional utility and plant patent applications filed after November 29, 2000, are published at this 18-month date. Since the examination of a new application typically takes longer than 18 months, most issued patents will have already been published before their issue dates. The issued patent may differ from the published application if amendments to the application have been made during examination. Publication may not occur at all: the inventor may choose to withdraw the application prior to its publication, or publication can be withheld at the request of the inventor under limited circumstances, or at the direction of the USPTO for reasons related to national security. Applications that are not published and do not become patents remain withheld from public access and do not become prior art.

[4] "Effective filing date" is defined in Section 2.4 of this chapter.

obtained from the later-filing inventor or whether the subject matters of both were commonly owned. Also note that the inventor cannot rely on the inventor's own public disclosure to overcome the reference if that public disclosure was more than a year prior to the effective filing date, since by such a disclosure the inventor would be creating a self-imposed bar to patentability, discussed in Chapter 3.

Figure 2.1 illustrates the principles of this category of prior art by comparing each of several references to the claims of an application under examination (or the claims of a patent whose validity is challenged). The horizontal arrows are time lines progressing from left to right, and for the application under examination (the time line at the top of the figure), the date of invention and the effective filing date are shown, while for each reference, the effective filing date and the date of publication

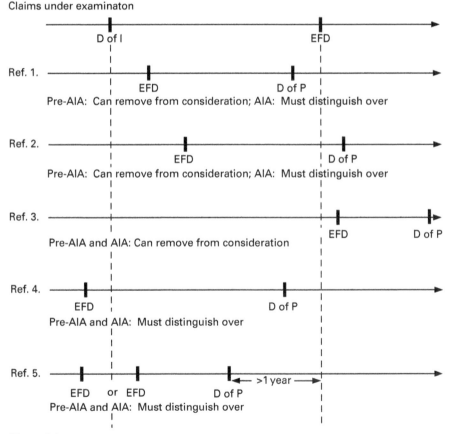

Figure 2.1   U.S. patents and published patent applications as prior art. "D of I," date of invention; "EFD," effective filing date; "D of P," either date of publication or date of patent, whichever is first. Choice between "Pre-AIA" and "AIA" is determined by the effective filing date of the claims under examination. The exceptions for references whose subject matter is obtained from the inventor, references whose subject matter was previously publicly disclosed by the inventor, and references that are commonly owned with the claims under examination are not shown, but would apply to Refs. 1, 2, and 4.

or patenting are shown. The various references differ in their effective filing dates, their issuance or publication dates, or both, all relative to the dates of the claims under examination. The text under the time line for each reference states the means by which the inventor can address the reference to establish the patentability of the claims under examination, and the results pursuant to both pre-AIA law and the AIA are presented. As noted in the caption, the distinction between pre-AIA and AIA is determined by the effective filing date of the claims under examination, not that of the reference. The chart shows that in certain cases, the reference can be removed from consideration as prior art, either by showing that the claims under examination had an earlier invention date (Refs. 1 and 2) or an earlier effective filing date (Ref. 3), while in other cases, the reference cannot be removed and the inventor must distinguish the claims under examination by showing how the subject matter of the claims differs from that of the reference. Note that the inventor claiming an earlier effective filing date may still have to show that the earlier filing date is indeed "effective," that is, that the application that was filed on that date provides proper support for the claims under examination.

2.3.2 Patents and Patent Applications Other than those of the United States

Searches for prior art through patent databases will often produce patents and published patent applications of other countries and authorities in addition to those of the United States. The Patent Cooperation Treaty (PCT) is one such authority, and international patent applications filed and published under the treaty are common examples of non-U.S. documents produced by these searches. The European Patent Office (EPO) is another such authority, producing published European patent applications and granted European patents, and further authorities are other regional and national patent offices throughout the world that variously publish regional and national patents as well as the applications themselves and abstracts of the applications. Since each of these documents is a publication that is accessible worldwide, each is as much prior art for its descriptive content as a U.S. patent or published U.S. application. This is true regardless of whether the claims of these documents are any narrower in scope than the descriptive material in their texts and, in the case of published applications and abstracts, whether they ever mature into patents. In addition, the exceptions applicable to U.S. patents and published patent applications, as set forth in Section 2.3.1, earlier in time apply to non-U.S. patents and published patent applications in the same way.

The difference between these non-U.S. patents and published applications and those of the U.S. when serving as prior art lies in their effective dates as prior art—the effective filing dates of the non-U.S. documents are their dates of publication rather than their filing dates. Non-U.S. patents and patent applications therefore lack the date dichotomy of their U.S. counterparts; the date on which these non-U.S. documents emerge as prior art against a U.S. patent application and the date that an applicant in the USPTO must predate with evidence of earlier action in order to remove these documents from consideration as prior art are the same date. For inventions governed by pre-AIA law, an earlier date of invention can suffice as that earlier action, while for inventions that are subject to the AIA, the only earlier action that will suffice is an earlier effective U.S. filing date. Whichever earlier action the U.S. applicant is entitled to use, the U.S. applicant need not be concerned with the filing date of the non-U.S. patent or published application, only its publication date, and since many of these non-U.S. documents are published at different stages, its earliest publication date.

Figure 2.2 illustrates the principles of this category of prior art in a manner analogous to Figure 2.1. Note that in all cases, the effective filing date of the reference is irrelevant.

Many of these non-U.S. patents and published patent applications will have U.S. counterparts, however. This occurs when the same application is filed in multiple countries or jurisdictions at different dates under procedures that allow the date of the

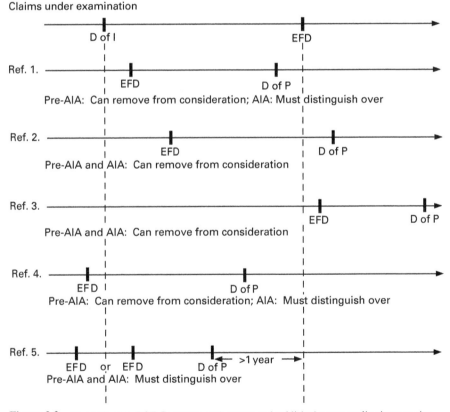

Figure 2.2   Non-U.S. (e.g., PCT, European, etc.) patents and published patent applications as prior art. "D of I," date of invention; "EFD," effective filing date; "D of P," either date of publication or date of patent, whichever is first. Choice between "Pre-AIA" and "AIA" is determined by the effective filing date of the claims under examination. The exceptions for references whose subject matter is obtained from the inventor, references whose subject matter was previously publicly disclosed by the inventor, and references that are commonly owned with the claims under examination are not shown, but would apply to Refs. 1 and 4.

earliest such filing to serve as the effective filing date for the succeeding filings in the other countries. U.S. inventors who seek patent coverage beyond the United States typically file an application first in the United States, then abroad, often through the PCT, using a procedure that will afford the foreign application the benefit of the filing date of the U.S. application. Foreign nationals commonly follow this procedure in reverse, by first filing a patent application in their native country or at a PCT receiving office or the EPO and later filing a U.S. patent application early enough to provide the U.S. application with the benefit of the filing date of the first-filed application. The filing date of the first-filed application then achieves significance as the effective filing date of the U.S. application. The prior art of concern is then the U.S. application rather than the foreign document, since the U.S. application, and not the foreign document, will have the benefit of the foreign document's filing date as its effective filing date. This will be true even though all of the documents are likely to be identical in their descriptive content. Thus, a reference PCT or European document that originated at the PCT or the EPO and was *not* later filed as a U.S. application can be eliminated by an inventor who can show a filing date that precedes the *publication date* of the reference, while a reference PCT or European document that originated at the PCT or the EPO and *was* later filed as a U.S. application will require the later inventor to show a filing date that precedes the reference's PCT or European *filing date*, since the PCT or European filing date becomes the effective filing date of the U.S. application, even if the actual filing of the U.S. application occurred much later. If a non-U.S. patent document of concern is produced by a search for prior art, one should therefore look for any corresponding U.S. document to see if the filing date of the non-U.S. document affords the U.S. document an earlier effective date as prior art than both the filing date of the U.S. document and the publication date of the non-U.S. document.

2.3.3   Other Published Literature

Publications other than patents and patent applications, such as the technical journal articles and other examples listed previously, typically do not involve complex determinations of their effective dates as prior art, since in all cases, the effective date is simply the date of publication. Neither the date on which a manuscript is submitted to a trade or technical journal for publication nor the date on which it is accepted for publication, for example, is of consequence; the paper's effective date as prior art is the publication date of the issue in which the paper will appear. Questions can still arise however as to what constitutes publication and at what date publication occurs. An example of a publication that does not involve actual dissemination to the public is the acquisition by a university library of an academic dissertation such as a doctoral thesis. The publication date of the dissertation as prior art is the date on which the dissertation is archived in the library records, whether by card catalog or by electronic database, so that a user can be informed upon perusing the records that the document is present in the library and can access it for reading. "Publication" is thus defined by accessibility to the public rather than by actual distribution.

Under pre-AIA law, an inventor who is not the author of the publication can remove the publication from consideration as prior art by showing an earlier date of invention, provided that not more than 1 year had passed between the publication date and the inventor's effective filing date. Under the AIA, the inventor's earlier date of invention will not suffice; the inventor must show an earlier effective filing date. The three exceptions that apply in the case of prior art patents and published patent applications apply here as well—the inventor who cannot show an earlier effective filing date can still remove the publication from consideration as prior art if the inventor can show that the subject matter of the publication was obtained directly or indirectly from the inventor, or that the inventor or one obtaining the subject matter directly or indirectly from the inventor had publicly disclosed the subject matter first, or that the subject matter of the publication and that of the inventors' patent application were commonly owned. Here as well, however, if the publication has a publication date that predates the inventor's effective filing date by more than 1 year, none of the exceptions apply. If the publication is the inventor's own publication, it does not constitute prior art under either pre-AIA law or the AIA, provided that the publication occurred no more than 1 year before the inventor's effective filing date.

Figure 2.3 illustrates the principles of this category of prior art in a manner analogous to Figure 2.1 and Figure 2.2.

2.3.4 Commercial Activities

Commercial activities that constitute prior art do so regardless of whether they are also published and in some cases regardless of the degree to which they are explained or otherwise described to the industry or the public at large, if at all. The two commercial activities that the patent statute identifies as prior art are "public use" of an invention and the placement of an invention "on sale." "Public use" is the use of a process, substance, material, or piece of equipment, depending on the invention that the use is cited against, in a commercial or production setting, and "on sale" includes entering into commercial transactions for the performance of the process or the sale of the substance, material, or equipment, as well as offers for the transactions. Prior art "public uses" are not limited to those that generate income to the user, sales are not limited in terms of the income they generate (compensation for a sale may be other than monetary), and offers for sale are not limited to those that are accepted. For a use to be "public," it must be accessible to the public, that is, not kept confidential. For sales and offers for sale to qualify, the subject matter of the sale must be "ready for patenting," that is, far enough conceived that it needs only a simple reduction to practice (if it has not already been so) for implementation or use. Exceptions to public uses and sales are those that are primarily for purposes of experimentation, that is, assessment of the technical functionality (e.g., as opposed to market appeal) of what is being implemented or sold. Patent courts recognize that the distinction between commercial use and experimentation is often a matter of degree, however, and those who seek to deny the prior art status of a commercial use or sale by claiming it was only for experimental purposes do not often succeed.

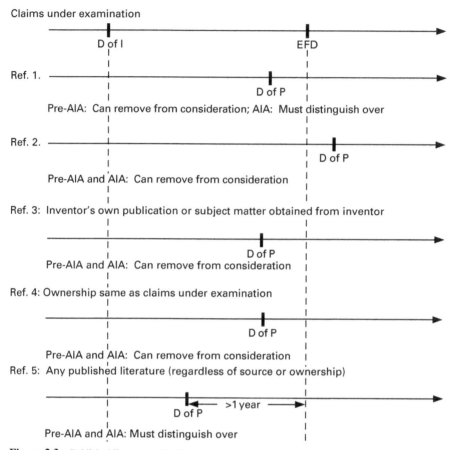

Figure 2.3　Published literature other than patents and published patent applications as prior art. "D of I," date of invention; "EFD," effective filing date; "D of P," date of publication. Choice between "Pre-AIA" and "AIA" is determined by the effective filing date of the claims under examination.

Under pre-AIA law, the only public uses and sales that constituted prior art were those that occurred in the United States; those confined to regions outside the United States were not prior art to the later-filing patent applicant. While public uses had to be accessible to the public to qualify as prior art, qualifying prior art sales and offers included those that were accessible to the public as well as those that were kept secret, such as sales made between parties who were under an obligation of confidentiality to the inventor. A patent applicant could remove public uses and sales (including offers) by others from consideration as prior art by showing that the inventor's date of invention was earlier, provided that not more than 1 year had passed between the use or sale date and the inventor's effective filing date, analogous to publication-type prior art. The AIA introduces several changes. The restriction to the United States is removed, and therefore public uses and sales occurring anywhere in the world can now qualify as prior art. The AIA introduces a new limitation however by eliminating secret sales from prior art. To remove public uses and sales from consideration as prior art, the first-to-file policy of the AIA requires the inventor to show an earlier effective filing date; an earlier date of invention by itself will not suffice.

Figure 2.4 and Figure 2.5 illustrate the principles of this category of prior art, Figure 2.4 addressing public uses and Figure 2.5 addressing sales and offers for sale. Each figure is presented in a manner analogous to the preceding figures, but a vertical

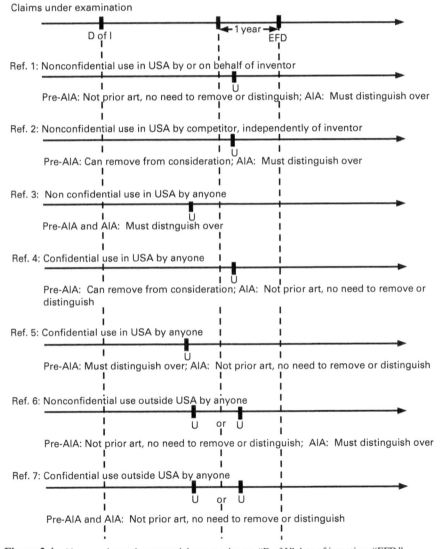

Figure 2.4   Nonexperimental commercial use as prior art. "D of I," date of invention; "EFD," effective filing date; "U," date of use. Choice between "Pre-AIA" and "AIA" is determined by the effective filing date of the claims under examination.

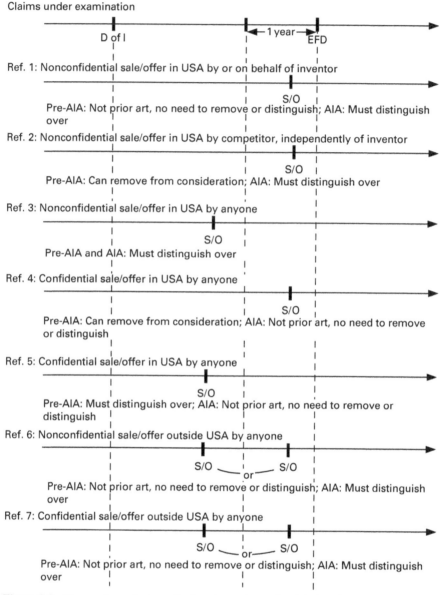

Figure 2.5 Nonexperimental sale or offer for sale as prior art. "D of I," date of invention; "EFD," effective filing date; "S/O," date of sale or offer. Choice between "Pre-AIA" and "AIA" is determined by the effective filing date of the claims under examination.

line has been added to indicate the 1-year date prior to the effective filing date of the claims under examination. Note that certain uses and sales or offers for sale are not prior art, and it is therefore unnecessary to either remove them from consideration as prior art or to distinguish over them in terms of subject matter. In some cases, this is due to the confidential nature of the use, sale, or offer and in others the fact that the use, sale, or offer is outside the United States.

2.3.5 Otherwise Available to the Public

Anything falling under this category that is not already covered by the published literature category is treated in the same way as published literature. The effective date of the prior art in this category is the date of availability, and the requirements for removal of items under this category as well as the 1-year limitation and the exceptions involving disclosure by the inventor or by one who obtained the subject matter either directly or indirectly from the inventor likewise apply.

2.4 A SERVANT OF TWO MASTERS?: THE "EFFECTIVE FILING DATE" AND ITS ROLE IN DETERMINING THE GOVERNING RULE

The term "effective filing date" is used in reference to U.S. patents and patent applications, both those under examination for the possible grant of a patent and those that are cited as prior art. As Sections 2.3.1–2.3.5 indicate, the term has relevance in both first-to-invent and first-to-file cases. Every patent application, published or otherwise, and every patent has an *actual filing date*, which is its date of receipt by the USPTO. The actual filing date may also be the *effective filing date*, but in many cases, the effective filing date is an earlier date, specifically that of another, earlier-filed patent application, published or not, which was still pending as of the filing date of the later application and of which the later application explicitly claims a filing date benefit. By the time the later application is either published or patented, the earlier application may be either withdrawn, abandoned, still pending, or itself matured into a patent, but in all cases, the use of its filing date as an effective filing date is unchanged. The earlier patent application may itself be a U.S. application but may also be a foreign or international patent application. Thus, a U.S. patent application may have the benefit of the filing date of an earlier-filed PCT patent application, an earlier-filed European (or other regional) application, an earlier-filed national application of a foreign country, or an earlier-filed U.S. application as its effective filing date. The latest-filed application may indeed be the last in a series of three or more applications, filed in succession, U.S. and otherwise, and enjoy the benefit of the first-filed of the series.

Applications with effective filing dates before March 16, 2013, will be governed by the pre-AIA examination standards including the first-to-invent rule, while those with effective filing dates on or after March 16, 2013, will be governed by the AIA examination standards and therefore the first-to-file rule. Individual claims of an application will be governed by the rule applicable to each claim, and an application with an actual filing date that is after the transition date but claims a filing date benefit from an application filed before the transition date may have certain claims that are entitled to the benefit and other claims that are not. Different claims may thus be governed by different rules.

Effective filing dates play a similar role in U.S. patents and published patent applications that are cited as prior art references. The prior art dates of these references are their effective filing dates under both standards. To defeat this prior art, therefore, the inventor whose patent is being examined can do so by showing an invention date that is earlier than the effective filing date of the reference if the later inventor's claims are entitled to the pre-AIA standard, but the inventor's *own* effective filing date must be earlier if the later inventor's claims fall under the AIA standard. As noted in the preceding section, non-U.S. patents and published applications are treated differently depending on whether the later inventor's claims fall under the pre-AIA standard or the AIA standard. In those cases where the filing date of the prior art patent or patent application is its effective date as prior art, the inventor against whose application the document is compared can only defeat the document by having an effective filing date that precedes the effective filing date of the document.

The use of the filing date of one application as an effective filing date of another, later application occurs in different ways, depending on whether the earlier application is a U.S. application or a non-U.S. application. The sharing of the **benefit of a filing date** across country lines is termed a "**right of priority**," established under the **Paris Convention for the Protection of Industrial Property**, first adopted in 1883. An application claiming a right of priority from an earlier application filed in a different country must be filed within 1 year of the earlier application and must make an express claim to the earlier application, listing its number, country, and filing date. International sharing of a filing date benefit can also be achieved by first filing an international application through the PCT and then filing the same application in one or more of the countries that are signatories to the PCT, a procedure referred to as "entering the national phase" of the PCT application, but one that is subject to time limitations set by the treaty.

The sharing of a filing date benefit between applications within the United States arises when an application is refiled after its original filing, either at the initiative of the inventor or pursuant to a requirement by the USPTO, in some cases followed by abandonment or withdrawal of the original application and in others without abandonment or withdrawal, leaving both applications pending or allowing both to mature into patents. One example of such a filing is a "**divisional**" **application**, which is filed in response to a "restriction requirement," that is, a decision by a USPTO examiner that the claims of an original application cover two or more inventions and thereby require separate searches. To respond to the requirement, the inventor must elect one of the inventions for examination and can then file one or more divisional applications as necessary to address each of the remaining (unelected) inventions. Another example is a "**continuation**" **application**, which is a refiling of the original application at the initiative of the inventor to allow the inventor more time and opportunity to respond to rejections from the examiner or to introduce a shift in claiming strategy. A third example is a "**continuation-in-part**" **application**, which is part continuation and partly an introduction of new descriptive material, possibly including new claims that may be broader in scope than those of the original. Continuation-in-part applications are often filed to incorporate useful material that has arisen after the filing of the original application.[5]

The benefit of an earlier U.S. filing date can also be drawn from a **provisional patent application**. Provisional patent applications are not actually applications for patent but rather documents that are filed with the USPTO as placeholders for true (nonprovisional) patent applications, that is, those that will be examined by the USPTO for possible grant of a patent. The purpose of a provisional application is thus to obtain a filing date that can be used as the effective filing date for a later-filed nonprovisional application. As Chapter 10 will explain in more detail, a provisional application can be filed without the organization or formatting features, or the full set of parts, that the USPTO requires of nonprovisional patent applications. Indeed, provisional patent applications are often documents prepared for other purposes, such as a publication in a technical journal, a presentation to potential investors, or a grant proposal. To obtain the benefit of the provisional's filing date, the invention must be made the subject of a nonprovisional application filed within a year of the provisional.

The value of a provisional application lies in its descriptive content; for a nonprovisional to have the benefit of the provisional's filing date, one or more of the claims of the nonprovisional must have descriptive support in the provisional. This in fact is true of any patent application that claims the filing date of an earlier patent application as its effective fling date, whether the earlier application is a U.S. application, provisional or nonprovisional, or a non-U.S. application, and whether the later application is itself a nonprovisional replacing a provisional, or a continuation, a continuation-in-part, a divisional, or a U.S. counterpart of a foreign application. Questions of the presence or adequacy of descriptive support arise most often when the earlier application is a provisional and the later is a nonprovisional or when the earlier is a nonprovisional and the later is a continuation-in-part. In both cases, the subject matter of certain claims of the later-filed application may be well described in the earlier application and therefore enjoy the benefit of the earlier filing date, while other claims lack that support and will therefore be limited to the later filing date.

2.5 CONCLUSION

How long will the pre-AIA law and the AIA coexist? Some sense of the answer can be gained from data on the pendency of patent applications together with the statutorily defined terms of issued patents. With certain exceptions, the statutory expiration date of a U.S. patent is 20 years from the patent's effective filing date. Notable

---

[5] In patent parlance, the terms "parent" and "child" are often used to describe the relation between U.S. patent applications that are either original and divisional, original and continuation, or original and continuation-in-part, the "parent" being the earlier filed of the pair. In cases involving multiple generations, the terms "grandparent," "grandchild," etc. are used, and the copendency requirement is satisfied by a chain of copendency, that is, at least one of the applications must be pending at all times between the first and the last.

exceptions are patents whose effective filing dates are drawn from a provisional application or a foreign priority application, in which cases the 20-year term is measured from the filing date of the U.S. nonprovisional application, thereby affording the U.S. patent an expiration date that can be as much as 21 years from the patent's effective filing date. When a patent is the subject of a lawsuit, the litigation can extend well beyond the patent's expiration date, in which case the applicability of pre-AIA law may do so likewise, extending the force of the pre-AIA law even further into the future. Pendency data from the USPTO indicate that the number of utility (nonprovisional) patent applications filed in the year 2012 was 542,815, increasing at a rate of about 40,000 applications per year, and as of May 2013, over 600,000 patent applications were pending in the USPTO and awaiting examination. The average pendency of a patent application as of May 2013 was approximately 30 months, but the body of pending applications is back-loaded due to a surge in both provisional and nonprovisional filings that occurred just before March 16, 2013, by those seeking to ensure that their applications fell under the first-to-invent rule. In the first 2 months of 2013, the number of patent applications filed per day was approximately 1,500, of which approximately 800 were provisional applications, while in the days immediately preceding March 16, the filing rate reached approximately 13,800 per day, of which approximately 13,000 were provisionals.

For these reasons, many pending applications will continue to be examined, and many issued patents will have their validity governed, by pre-AIA standards well beyond the transition date. It will therefore be necessary to maintain familiarity with both the pre-AIA and AIA systems well into the year 2034 and possibly beyond. A certain degree of comfort however can be drawn from the fact that the greater number of cases where pre-AIA law will apply will be pending patent applications undergoing the process of examination. With the current average pendency of 30 months and the stated goal of the USPTO to reduce this to 20 months, an early tapering off of the number of cases subject to the pre-AIA standard can be expected.

Chapter 3

Creating One's Own Prior Art: Self-Imposed Barriers to Patentability

It's a safe assumption that most individuals who patent their inventions proceed by inventing first and then applying for a patent rather than, for example, conceiving of the invention as they sit at a computer keyboard drafting a patent application. For most inventors, therefore, the primary focus is on the creative process of inventing, the problem solving that leads to the invention, or the use of their imaginations to explore new frontiers. The patenting process is secondary for these inventors in terms of their time and attention, their career advancement, and often their financial resources. The inventor engaged in an ongoing business concern, for example, will rightfully focus on the operation of the business and on the retention and satisfaction of customers in addition to drawing in new customers. The researcher in an academic environment will focus on scientific advances and on enhancing his or her reputation in the academic and scientific communities. The result of placing one's primary focus on the creative process is that inventors, and particularly those who do not already have patents, tend to take certain actions before their patent applications are filed that can jeopardize their rights to a patent or at least that impose time limits on their ability to file their applications. Inventors may be tempted to think that their patent applications are immune from their own prior actions, particularly with the first-to-file standard that suggests that obstacles to patenting arise mostly from the filing of competing patent applications by others rather than what one does oneself prior to filing. The truth however is that certain areas under the definition of prior art, including those implemented by the AIA, place the inventor's own acts on an equal footing with those of the inventor's competitors and others in the technological community.

The unwary inventor may thus create his or her own prior art and in so doing may block or severely limit the inventor's patent rights. From a practical standpoint, inventors cannot entirely avoid the actions that pose these risks, given the demands

First to File: Patents for Today's Scientist and Engineer, First Edition. M. Henry Heines.
© 2014 the American Institute of Chemical Engineers, Inc. Published 2014 by John Wiley & Sons, Inc.

and interests of running a business or of advancing one's career or one's professional standing. Nevertheless, with an understanding of the most common ways in which inventors unwittingly create their own prior art, the risks can be controlled and even avoided entirely, and potential patent rights can be preserved. This chapter addresses two prime areas where practices that inventors commonly engage in prior to filing a patent application raise the risk of creating prior art. The first is the placement of the subject matter of the invention on sale, and the second is the publication of the subject matter.

3.1   THE ON-SALE BAR

As noted in Chapter 2, a sale or offer for sale of the subject matter of an invention that is sufficiently developed or at least sufficiently thought out that it is ready for patenting and that occurs more than 1 year before the effective filing date of a patent application claiming the invention will bar the application from becoming a valid patent. Such a sale or offer is thus referred to as an "**on-sale bar**." If the sale or offer occurs after March 16, 2013, it can qualify as a bar regardless of the part of the world where it is made, but sales or offers whose terms require the parties to keep the subject matter confidential will not qualify. Since the present chapter seeks to advise inventors regarding their own present or prospective commercial activities, the relevant parameters of the on-sale bar are those that incorporate the modifications introduced by the AIA. With this in mind, the key features of the on-sale bar with respect to an inventor's own activity are its exceptions, that is, sales or offers that do *not* create an on-sale bar. These are secret sales or offers, sales occurring or offers made 1 year or less before the inventor's effective filing date (the date 1 year before the effective filing date being the "critical date"[1]), and sales or offers whose purpose is experimental with regard to the functional viability of the subject matter sold or offered.

3.1.1   Ready for Patenting?

An invention is clearly "ready for patenting" when it has been reduced to practice, that is, when it or a prototype has been constructed or otherwise implemented (such as in the case of a process), and been shown to function for the purpose for which it was designed. Can it also be "ready for patenting" before it has been reduced to practice?

The question came up in connection with U.S. Patent No. 4,491,377,[2] which was in the field of chip packaging for integrated circuit dies. The particular area of interest was high-density packaging for large-scale integration, that is, chips that contained

---

[1] Certain commentators refer to the effective filing date itself as the "critical date," but most, and more correctly so, use the term to refer to the date 1 year prior to the effective filing date, as we do here.

[2] "Mounting Housing for Leadless Chip Carrier," United States Patent No. 4,491,377; Pfaff, inventor; issued January 1, 1985.

a high density of circuit functions per chip and the sealed enclosures that protected these chips against damage from impact or the environment while allowing the chip circuitry to be connected to external components such as a printed circuit board. Because of its large number of circuit functions, a chip of this type required an enclosure ("package") with a large number of leads, and to accommodate so many leads in a single package, individual leads had to be unusually long. This created high impedance in the package, which shortened the useful life of the package and in some cases made the package inoperative. To address this problem, "leadless" chip packages (or "carriers") had been developed that had terminal lands in place of the leads, the lands making their electrical connection with the elements on a printed circuit board by simple contact. For functioning electrical connections, of course, it was important for the lands to be properly aligned with the elements on the board. Unfortunately, the arrangement of the elements on the board varied from one board to the next, and no standard had been established for achieving the proper alignment. Instead, individual circuit boards were constructed with special mounting housings to receive the carriers. These housings however required precise registration of the carrier with the housing, and while this could be achieved by the geometry of the carrier, the ceramic base of the typical carrier could not be machined to enough precision to assure proper and consistent registration.

The solution provided by the patent was a special mounting housing (or "socket") that is shown in Figure 3.1A, B, and C, which are drawings taken from the patent.

Figure 3.1A is a top view of a leadless chip carrier (10) (the square element in the center of the drawing; the chip itself is embedded in the carrier and is not visible) and the socket (the entire structure surrounding the carrier), and Figure 3.1B is a vertical cross section of the carrier and socket with the carrier again indicated by the number 10. As seen in Figure 3.1B, the socket is constructed in three parts—a base support (15), an intermediate support (16), and a spreader cap (40). The intermediate support is rigidly secured to the base support, while the spreader cap is mounted on posts that allow it to move up and down above the intermediate support. Secured between the base support and the intermediate support are a series of "pins" (20). Each pin extends from a pin shank (20d) at the bottom, protruding through the base support with its end exposed for connection to a printed circuit board, to a pair of blades (20a, 20b) at the top, extending into the interior of the spreading cap. One pin is included for each terminal land, and the inner blade (20b) of the pin makes electrical contact with its corresponding terminal land. The blades are resilient and have sloping edges at their upper ends. The spreader cap (40) is open at its center to allow insertion and removal of the chip carrier (10), and a downwardly projecting "lip" (43) runs along the inner rim of the cap surrounding the open center. The lip engages the sloping edge of the inner blade (20b) at the top of the pin, forcing the blade outward as the cap is pressed down, as shown in Figure 3.1C. To place the carrier in the socket, the cap is pressed down, spreading the inner blades of all of the pins to clear the opening for the carrier. Once the carrier is inserted, the cap is released, and the resilient blades spring back to their upright position, with their inner edges making electrical contact with the lands.

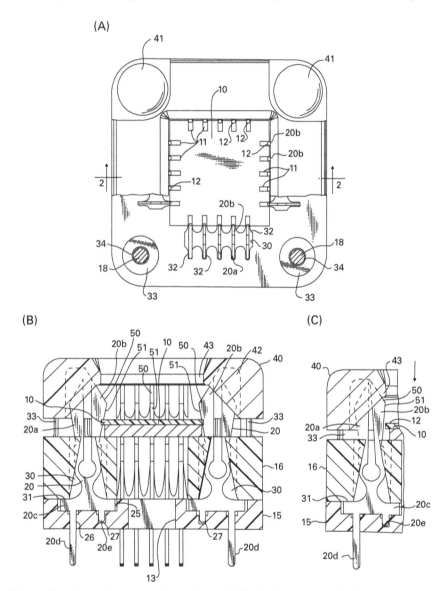

Figure 3.1   Selected figures from Patent No. 4,491,377. (A) Top plan view. (B) Sectional view through lines 2–2 of (A). (C) Partial sectional view through lines 2–2 of (A).

More than a year before the patent application was filed, that is, before its critical date, the inventor had prepared detailed engineering drawings of the socket that showed the design, dimensions, and materials of construction and sent them to a company called Weiss-Aug for customized tooling and production. At about the same time, and also before the critical date, the inventor also prepared a sketch that he sent to Texas Instruments who responded with a purchase order. No prototype had yet been made, and since Weiss-Aug needed time to develop the customized tooling for the units, Weiss-Aug was not able to produce the units until several months later, after the critical date. Once the patent issued, the inventor sued Wells Electronics for patent infringement.

When Wells learned of the drawings prepared for Weiss-Aug and the purchase order from Texas Instruments, Wells raised the defense that the patent was invalid because the purchase order constituted placement of the invention on sale before the patent's critical date. The inventor responded that since there was no physical embodiment of the invention at the time of the purchase order, the order did not constitute a placement "on sale" within the meaning of the patent statute, and the trial court agreed. The appeals courts, including the Supreme Court,[3] disagreed, stating the invention was indeed "ready for patenting" at the time of the order despite the lack of a prototype. The Court reasoned that the primary meaning of the word "invention" in the statute referred to the inventor's conception of an idea rather than to a physical embodiment of that idea and that an invention could indeed be patented before it had been reduced to practice. While a concept had to be complete before it could be an "invention" and reduction to practice was the best evidence of completion, reduction to practice was not necessary in every case. Indeed, the drawings and other descriptions of the invention that the inventor had prepared were sufficiently specific to enable a person skilled in the art to practice the invention, and therefore, the invention was on sale at the time of the purchase order, setting the start of the 1-year time limitation that the inventor failed to meet.

Can an on-sale bar result from an offer for sale that was made before the conception of the invention was complete?

Like the Pfaff case, the invention in August Technology Corporation's U.S. Patent No. 6,826,298[4] arose in the semiconductor industry, but its focus was on inspection systems for detecting defects in semiconductor wafers. The patent states that prior to the invention, wafers were typically inspected by a series of optical inspections at various stages throughout the fabrication of the wafer and after the wafer was packaged, the inspections being performed at scales beginning at 0.1–1 micron at the masking and circuitry formation stages and progressing to 5 microns and above after packaging. Automated equipment for performing the smaller-scale inspections was available but costly and therefore not economically justifiable for wafer fabricators other than those producing wafers in high volume. Even for high-volume fabricators, there was a lack of equipment for performing the intermediate-level inspections (at the 1-micron scale), which were particularly important since these inspections were done on the fully formed but as yet unpackaged wafers to check for defects in metallization, diffusion, passivation, and other irregularities that might affect the functioning of the circuitry. These intermediate inspections were performed by individual operators using microscopes, which introduced human

---

[3] *Pfaff v. Wells Electronics, Inc.*, 525 U.S. 55 (1998).

[4] "Automated Wafer Defect Inspection System and a Process of Performing Such Inspection," United States Patent No. 6,826,298, O'Dell et al., inventors; August Technology Corporation, assignee; issued November 30, 2004.

error. August Technology addressed these problems by developing an automated system for the fully formed wafers that operated by first forming images of wafers that were known to be properly fabricated and defect-free (i.e., "good" wafers) using strobe lighting and a scanning camera to generate pixels, assembling the pixels into full images, analyzing the images to develop a model set of parameters from the good wafers, then scanning and imaging the test wafers in the same way, and finally comparing the parameters of the test wafers to those of the model set. Each step was controlled and operated by software, including coordinating the strobe light with the camera, collecting the pixels and assembling them into the image, storing the model set of parameters, comparing the parameters of the test wafers to those of the model set, displaying the images and indicating the detected defects, and performing all mathematical and statistical functions necessary for filtering and sensitivity to achieve the desired resolution. When the patent was granted, its owner August Technology sued competitor Camtek, Ltd., and among the defenses that Camtek raised was a prior art defense based on an offer for sale that August Technology had allegedly made more than a year before the patent's critical date.

The offer for sale that Camtek was pointing to was a transaction between August Technology and a wafer fabricator named ICS, Inc. ICS first approached August Technology with a request to develop an automated wafer inspection machine to meet ICS's special needs. In response to the request, August Technology issued ICS a purchase order for the machine, the purchase order calling for an initial payment by ICS of 15% of the purchase price and additional payments at various stages with a final payment once the unit had been shipped to and accepted by ICS. Soon after the purchase order was issued, ICS made the initial payment and August Technology began its preliminary hardware design for the machine. August Technology took several months to complete its hardware and software design, and a year after the purchase order, a fully designed and manufactured unit was shipped to ICS for its inspection. Nine months after that, August Technology filed a provisional patent application, whose filing date became the effective filing date of the patent. The purchase order had thus been issued before the critical date, and at the time the purchase order was issued, the intention was still in its early stages of conception. Shipment of the unit, on the other hand, did not occur until after the critical date.

The trial court considered the evidence and concluded that the purchase order did not constitute an on-sale bar even though it occurred before the critical date. The court reasoned that without completed designs for either the software or the hardware, the unit was not ready for patenting at the time of the purchase order. Camtek challenged this on appeal, arguing that the condition of the invention at the time of the purchase order did not tell the full story and that if conception had been completed at some point between the date of the purchase order and the critical date, the invention would still have been subject to the on-sale bar. The appellate court agreed[5] and explained that if an offer for sale is *made and retracted* prior to the completion of the conception, there has been no offer for sale within the definition of prior art, whereas if the offer *remains open*, a subsequent completion of the conception causes the offer

---

[5] *August Technology Corporation et al. v. Camtek, Ltd.*, 655 F.3d 1278 (Fed. Cir. 2011).

to be prior art as of the date the conception became complete. The August Technology purchase order had not been retracted, and since the evidence before the trial court did not indicate exactly when the hardware and software had been sufficiently developed that a functioning unit could be built, the appellate court held that the trial court's conclusion had been based on insufficient evidence. The case was ultimately decided on other grounds, but the rule stated by the appellate court for an on-sale bar was that an invention that is offered for sale before it is actually invented can still raise the on-sale bar provided that invention actually occurs before the critical date.

A series of sales that were never consummated and a series of rejected prototypes were the subject of a case regarding Patent No. 6,834,603.[6] The patent was concerned with commercial sewing machines, and particularly those for assembling pillow-top mattresses. Figures from the patent are reproduced as Figure 3.2A and B.

The focus of the invention was on the gusset (110 of Fig. 3.2A), which is the elongated strip of fabric that extends along the periphery of the mattress to join the padded layer (108, the "pillow top") to the panel (106) that forms the top of the mattress body (102). The gusset is folded along its centerline and oriented such that the fold line is recessed from the outer edges of the pillow top and the panel while one outer edge of the gusset is sewn to the outer edge of the pillow top and the other

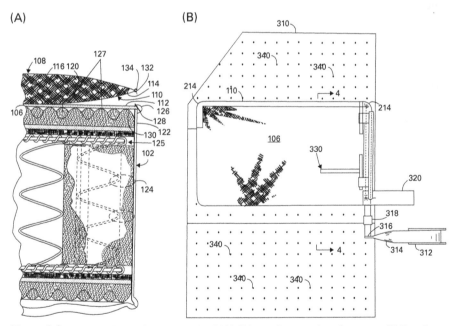

Figure 3.2   Selected figures from Patent No. 6,834,603. (A) Cross section of mattress. (B) Top plan view of gusset manufacturing machine.

[6] "Attachment Gusset with Ruffled Corners and System for Automated Manufacture of Same," United States Patent No. 6,834,603, Price et al., inventors; Atlanta Attachment Company, assignee; issued December 28, 2004.

to the outer edge of the panel. The gusset is also mitered (pleated) at each of the four corners of the panel to provide a smooth contour at each corner and to help secure the pillow top to the panel around the entire periphery of the mattress. The background description in the patent states that gussets of this type were traditionally formed by hand, starting with forming the individual legs of the gusset, then sewing the legs together at the mitered corners, all before sewing the gusset to the panel, and that this was a tedious and expensive process in need of simplification. The inventors at Atlanta Attachment Company, the owner of the patent, addressed this need by designing an automated system that included a sewing table (310 of Fig. 3.2B) with a turning mechanism (330) for holding the mattress panel and a sewing machine (320) positioned above the table, a supply reel (312) for the gusset material, a gusset guide and folding mechanism (316) for directing the gusset material to the sewing machine and folding it as it was being sewn, an edge detector (not shown) for detecting when a corner of the panel approached the sewing machine, and a pleat generator (not shown) actuated by the edge detector. The system also included various guides, pneumatic actuators, and a conveyor. A provisional patent application was filed, and its filing date ultimately became the effective filing date of the patent, defining the critical date as 1 year prior to the filing date of the provisional.

The patent contained 48 claims, variously expressing the invention as an apparatus, a method, and a system and differing widely in their breadth of coverage. The leading system claim listed five components—a gusset forming station for forming the gusset from a strip of material, the sewing table, the sewing machine, the pleat generator, and a system controller for running the sewing machine at a high sewing rate when straight portions of the gusset were passing through it and a lower sewing rate for the mitered corners. Other system claims added in further features, including clamp arms for holding the mattress body panel and turning it as the corners reached the sewing machine, air jets in the surface of the sewing table to provide an air cushion for the panel, a gusset accumulator for the gusset station, and a gusset folder.

The prototypes were constructed in response to a request by Sealy, Inc., for Atlanta Attachment to create an automatic gusset ruffler machine, and each prototype was presented to Sealy for sale soon after it was made. The first three prototypes were all sent before the critical date. Sealy tested them and found that they required too much operator control, first to determine when to create the pleats and then to create a ruffle, and both prototypes were returned to Atlanta Attachment without payment. The problems were corrected in the third prototype, which Sealy paid for. The third prototype also failed to satisfy Sealy however since it vibrated at high-speed operation and since its pleat generator was only able to form one stitch per pleat. Atlanta Attachment responded by refunding the amount Sealy had paid for the third prototype and presented Sealy with a fourth, this time after the critical date. The fourth prototype corrected the vibration problem by substituting a pneumatic drive for the eccentric drive in the pleat generator of the first three prototypes and also corrected the one-stitch-per-pleat limitation by decoupling the pleat generator from the sewing head. Nevertheless, Sealy ultimately decided not to purchase machines from Atlanta Attachment.

After the patent issued, Atlanta Attachment sued one of Sealy's competitors, Leggett & Platt, Incorporated, claiming infringement. Leggett & Platt raised several defenses, one of which was that the prototypes that Atlanta Attachment had presented to Sealy before the critical date amounted to a placement of the subject matter of the claim on sale. Atlanta Attachment argued that the problems in the three precritical date prototypes and its ongoing efforts to solve the problems prevented the invention from being considered "on sale" within the meaning of the on-sale bar. The trial court agreed, but the appellate court disagreed,[7] noting that neither vibration-free operation nor an uncoupled pleat generator was recited in the leading system claim as features of the invention and that therefore the changes introduced after the third prototype were refinements that did not preclude reduction to practice of the invention as recited in the claim. The defects that the changes corrected would only make the claim invalid if they prevented the invention as claimed from being workable or useful.

The Atlanta Attachment case included both an offer for sale (the purchase order) and the shipment of prototypes before the critical date, all with development of the invention still in progress, and yet each was held to be an on-sale bar. An offer can thus have "on-sale" status without any exchange of funds, and even if the offer, which included production models in addition to the prototypes, is ultimately rejected, as it had been by Sealy.

It is interesting to note that the prototypes and the communications between Atlanta Attachment and Sealy were kept confidential by agreement between the parties, but this had no bearing on the outcome of the case. The reason was that the patent predated the America Invents Act (AIA), and therefore, the confidentiality of the transaction was of no consequence to the question of whether it qualified as a placement on sale. With the enactment of the AIA, secret sales do not qualify as on-sale prior art, and this provision is effective only prospectively. Therefore, if the patent had had an effective filing date of March 16, 2013, or later, the presentation of three earlier prototypes would not have given rise to an on-sale bar.[8]

3.1.2 Exceptions for Experimental Use

Separately from its argument of not being "ready for sale," Atlanta Attachment also argued that neither the purchase order nor the sales of the first three prototypes constituted an on-sale bar because the prototypes were for experimental purposes. The court rejected this argument as well, observing that the purposes of the testing by Sealy and the refinements introduced by Atlanta Attachment with each successive prototype were to produce a machine that satisfied Sealy's requirements rather than simply a machine that was suitable for the purpose of the invention as stated in the patent. Admittedly, each successive prototype functioned better than its predecessor, but the

---

[7] *Atlanta Attachment Company v. Leggett & Platt, Incorporated*, 516 F.3d 1361 (Fed. Cir. 2008).

[8] With the distinction between secret and nonsecret sales having been introduced so recently, case histories exploring what is considered secret and what is not are likely to occur in the coming years.

three prototypes that predated the critical date still succeeded in attaching the gusset to the mattress panel, which, according to the patent, was the purpose of the invention in its simplest form.[9] The court also observed that regardless of the reasons for Sealy's dissatisfaction with the early prototypes, the testing performed at Sealy's facilities was done independently, that is, without any retention of control by Atlanta Attachment. The **experimental use exception** applies only to actions of the inventors and their agents, so the lack of control by Atlanta Attachment over Sealy's testing was a further reason why the sales of the prototypes did not qualify for the exception.

Even if the testing is performed by the inventor, the testing may still fail to qualify for the experimental use exception. An example is found in a case involving Patent No. 4,492,779,[10] which relates to rocket motor construction, particularly the construction of rocket motor casings. The casings tended to deteriorate when exposed to the high temperatures encountered in flight and when eroded by contact with droplets of aluminum oxide in the combustion gas generated inside the motor. The deterioration problem had previously been solved by applying an asbestos-containing coating to the casing wall, but the environmental and health concerns associated with asbestos prompted rocket manufacturers to seek alternate insulation materials. Thiokol Corporation (its name at the time) responded by developing an elastomeric composition that was a mixture of aramid polymer fibers and any of various inorganic powders and proceeded to apply for and obtain the patent. To show the utility of the invention, the patent contained erosion data obtained during a static test performed on a motor casing insulated with the novel composition. The test consisted of burning rocket propellant inside the motor casing for 5 s under conditions of interior pressure and exhaust velocity that simulated the environment of the motor casing in flight.

Well before the filing date of the patent (which was also its effective filing date), Thiokol was awarded a contract by the U.S. Air Force to develop and deliver rocket motors with insulation that was free of asbestos. The contract specified the characteristics of the motors and the conditions of delivery and called for tests to be performed in two phases, some by Thiokol and others by the Air Force. Thiokol's phase I tests included design verification tests and an accelerated aging study, while those of the Air Force were cook-off tests to determine whether the insulation can withstand a fire and prevent the fire from reaching the propellant inside the motor. Phase II testing was to begin after the motors passed a critical design review and included preliminary

---

[9] The refinements that were introduced during the testing in the Atlanta Attachment case actually appeared in certain claims of the patent, but not in the leading system claim, and their absence from that claim was the reason that the on-sale bar defense was raised. If only those claims that listed the refinements were at issue, the court's conclusion may have been different. The reason that the broader claim (lacking the refinements) raised the issue is that the prototypes were not brought to the attention of the patent examiner to allow the examiner to assess them as possible prior art against the broader claim. This also raised the possibility of "inequitable conduct" by Atlanta Attachment before the PTO, which, had it been found, would have infected the validity of the entire patent.

[10] "Aramid Polymer and Powder Filler Reinforced Elastomeric Composition for Use as a Rocket Motor Insulation," United States Patent No. 4,492,779; Junior et al., inventors; Thiokol Corporation, assignee; issued January 8, 1985.

flight rating/qualifications by Thiokol and flight, service life, and hazard classification tests by the Air Force. The Air Force paid Thiokol for the motors to be used in the testing.

The date that Thiokol shipped the first motors for testing was before the critical date of the patent application, and while Thiokol had completed the design verification tests by then, the accelerated aging study was still in progress and continued through the critical date. Once the patent issued, Thiokol sued its competitor Alliant Techsystems, Inc., for patent infringement. Alliant responded by raising Thiokol's sale of the rocket motors to the Air Force before the critical date as a defense, and Thiokol responded by arguing that since the accelerated aging study had not been completed before the critical date and was a requirement of the contract, the sale was for an experimental purpose and thus qualified for the experimental use exception. The court ruled against Thiokol on this issue,[11] explaining that for testing to qualify for the exception, the tested feature must be a claimed or inherent feature of the patented invention. The aging study tested durability rather than insulating ability, and since no mention of durability appeared anywhere in the patent, the experimental use exception did not apply, and the patent was held to be invalid due to the on-sale bar.

3.2   THE PUBLICATION BAR: PUBLISH *AND* PERISH?

The "**publication bar**" is the barring by the patent statute of a patent application on an invention that had been described in a printed publication more than a year before the effective filing date of the application. The relevant section of the statute also applies to publications that do not reveal the invention itself but do provide information close enough to the invention to raise questions of the invention's novelty and non-obviousness, in which case the publication has the status of prior art. In both cases, the statutory section applies regardless of whether the publication is authored by the inventor(s) or by someone else. In this chapter, our focus is on those publications by the inventor(s) or an agent of the inventor(s), such as an employer or an inventor's assistant, and particularly those publications that reveal the invention itself, since these tend to occur among individuals who are unaware of how widely the term "printed publication" has been interpreted and how easily and unintentionally they can destroy their own patent rights as a result.

For a publication to constitute a bar to patentability, the publication must be sufficiently descriptive that a person skilled in the art, that is, one who despite being previously uninformed about the invention is sufficiently well versed in the technology to follow the inventor's instructions, will be able to reproduce the invention upon reading the information contained in the publication. This requirement has its parallel in the "ready for patenting" requirement of the on-sale bar but is easier to establish since the adequacy of the description can typically be determined within the confines of the publication itself as read by the person skilled in the art. Also, there are no exceptions to the publication bar for "experimental use." Other issues arise,

---

[11] *Cordant Technology et al. v. Alliant Techsystems et al.*, 45 F.Supp. 2d 398 (D. Del. 1999).

however, notably the question of whether a "printed publication" within the meaning of the statute has actually occurred, and if so, when.

The expression "printed publication" first appeared in the Patent Act of 1836 and has remained in the statute ever since, even with the modifications introduced by the AIA.[12] In the decades and centuries since 1836, of course, means of publication and the dissemination of information have evolved and changed radically, from the printing press (in existence in 1836) through the typewriter (introduced in the 1860s), then copy machines, fax machines, microfilm, electronic recordation, and finally posting on the Internet. While the wording in the statute remained unchanged, the patent courts have had to deal with the fact that many of the contemporary forms of publication do not involve "printing" in the conventional sense and public accessibility to information and the dissemination of information are now achieved in ways never contemplated in 1836. Courts have therefore had to adjust their definition of a "printed publication" to recognize and adapt to each new technology, including the new means of data storage, retrieval, and dissemination. The courts have done this reluctantly, at first denying the status of prior art to each new form of publication as it first arose in view of its limited use. As the use of each new form became widespread, the courts eventually agreed that the forms did qualify, although with a different set of criteria for each form. Here, we address the more common forms of prefiling "printed publication" that inventors tend to engage in.

3.2.1 Posting on an Internet Server

An Internet server is not a printer, nor does it disseminate files that are stored on it. Instead, a server, particularly a File Transfer Protocol (FTP) server, functions primarily as a repository for the files from which the files can be retrieved by an electronic network, notably the Internet. Once retrieved, the files can be copied to a personal computer or simply displayed on a monitor connected to a personal computer, or both, in either case allowing them to be printed or simply read without printing. Regardless of what the user does with a retrieved file, however, the posting on the server can qualify as a "printed publication."

The issue was addressed directly in an infringement suit brought by SRI International, Inc., against Internet Security Systems, Inc., involving four SRI patents,[13] all on inventions relating to cybersecurity and intrusion detection. In more descriptive terms, the invention in these patents was a "computer-automated method of hierarchical event monitoring and analysis within an enterprise network," that is, a tool for tracking malicious activity across large Internet networks. All four patents claimed filing date benefits from a single patent application that was filed on

---

[12] 35 U.S.C. 102(a)(1), per the AIA.

[13] "Network Surveillance," United States Patent No. 6,321,338, issued November 20, 2001; "Hierarchical Event Monitoring and Analysis," United States Patent No. 6,484,203, issued November 19, 2002; "Network Surveillance," United States Patent No. 6,708,212, issued March 16, 2004; and "Network Surveillance," United States Patent No. 6,711,615, issued March 23, 2004—all naming Porras et al., as inventors and SRI International, Inc., as assignee.

November 9, 1998, resulting in a critical date of November 9, 1997. The inventors described the invention in a paper that SRI submitted to the Internet Society (ISOC) for presentation at ISOC's 1998 Symposium on Network and Distributed Systems, the presentation occurring after the critical date. The submission of the paper occurred much earlier, however, on August 1, 1997, in response to ISOC's call for papers. At the same time as they made their submission, the authors placed a backup copy of the paper on SRI's FTP server. The server was publicly available, but files on the server were accessible only to those who were given the files' full FTP address. Nevertheless, the server posting occurred before the critical date, and Internet Security Systems argued that the posting amounted to a precritical date publication invalidating the patents. The trial court agreed, but the decision was overturned on appeal.[14]

The inventors had provided four individuals with the full path and file name of the posted paper upon the individuals' requests. The court noted however that the posting was not cataloged or indexed in any meaningful way and not publicized or placed in front of the interested public other than the four individuals who requested it, and for this reason, the court held that the posting did not constitute a "printed publication." What is meant by "printed publication" in an Internet context, therefore, is accessibility to the interested public, which means something other than simply being posted on a server without any means of locating the file other than by those who possess the full path and file name.

3.2.2 Slide Presentations and Posters at a Conference

Another means by which researchers release the results of their research at symposia and trade shows are slides that the researchers use to supplement oral presentations. An oral presentation itself is clearly not a "printed publication," but the question is raised by the slides since although not "published" in the conventional sense, they are certainly displayed.

Inventors Klopfenstein and Brent were studying the use of soy cotyledon fiber (SCF) as a food additive, which was already known to reduce serum cholesterol and increase HDL cholesterol when the SCF was incorporated into food in extruded form. The inventors found that both the reduction in serum cholesterol and the increase in HDL cholesterol were improved to an unexpected degree when the SCF was subjected to a double extrusion, and this discovery was the subject of their patent application. Two years before the patent application was filed, however, and thus a year before the critical date, the inventors appeared at a meeting of the American Association of Cereal Chemists and presented their discovery to the meeting attendees, using a series of slides in their presentation. The slides were also printed and pasted onto poster boards where they were displayed for two and a half days.

In performing their duty of candor before the Patent and Trademark Office (PTO), the inventors reported the slide presentation and poster showing to the

---

[14] *SRI International, Inc., v. Internet Security Systems, Inc.*, 511 F.3d 1186 (Fed. Cir. 2008).

examiner but argued that since the slides and posters were neither distributed nor copied, they did not amount to a printed publication and therefore should not present a bar to their patent application. The examiner disagreed and rejected the application, and the inventors took their case first to the Board of Patent Appeals and Interferences (the next higher level in the PTO at that time) and finally to the Court of Appeals for the Federal Circuit. Both the Board and the Court agreed with the examiner.[15] The Court explained that the governing rule was not whether copies of the slides had been disseminated but rather whether the slides had been made publicly accessible. In this case, the slides had been shown to a wide variety of viewers who were educated in the field in which the subject matter lay, and the inventors had made no instructions to anyone seeing the slides or the posters that they were not to be copied or reproduced, all of which made them sufficiently accessible to the relevant public that they fell within the meaning of a "printed publication." The patent application was therefore refused, and no patent was granted on it.

3.2.3   Submission of a Thesis to a University Library

For inventors pursuing academic degrees in institutions of higher learning, a dissertation that is part of the degree requirement will likely be the main focus of the inventors' attention, second to the research itself. Since university libraries where the dissertations are received and archived will neither print nor distribute copies of the dissertation other than upon a request, the submission of a thesis would seem to present a difficult case for a "printed publication." Nevertheless, the courts have established a special set of rules for this fact situation, illustrated by a case involving Patent No. 4,225,672.[16] In this case, the thesis was not that of the inventor but instead one prepared by a different individual independently of the inventor. The governing rules however are the same.

   The patent was in the area of clinical assays for amylase activity, amylase being an enzyme in the human body that catalyzes the conversion of starch to various sugars. The level of amylase activity in the body is of clinical importance in the diagnosis and treatment of patients for diabetes and pancreatitis, as well as in research related to these diseases. The assays consisted of testing a sample of the patient's bodily fluid by contacting the sample with a substrate, that is, a substance that would indicate, particularly by changing color, whether the active enzyme was present in the fluid. The substrate addressed by the patent was a derivative of a maltooligosaccharide glycoside, which, before the invention, was obtained by organic synthesis, that is, the reaction of simpler starting materials with each other using common chemical reactions. Unfortunately, the yield of the reaction and the purity of the product it produced were so low that the substrate was costly to prepare. The invention lay in the discovery that one could obtain the maltooligosaccharide glycoside

---

[15] *In re* Klopfenstein, 380 F.3d 1345 (Fed. Cir. 2004).

[16] "Method for Producing Maltooligosaccharide Glycosides," United States Patent No. 4,225,672; Hall, inventor; issued September 30, 1980.

derivatives in substantially pure form by using another enzyme known as glucano-transferase to promote the reaction, this enzyme being readily isolated from a variety of biological sources. The patent was applied for on February 27, 1979, and granted on September 30, 1980. Several months after the grant of the patent, the inventor applied to the PTO for reissue of the patent to correct a clerical error in its grant.

Independently of the inventor, who was a resident of Alabama, a doctoral candidate at Freiburg University in Germany was studying the same reaction and made the same discovery of the effectiveness of glucanotransferase. The doctoral candidate described his research and its results, including the discovery, in a dissertation that he submitted to the university library in 1977, which was before the patent's critical date of February 27, 1978. Upon discovery of the patent and the fact that it was once again before the examiner in a **reissue application**, a competitor of the inventor who was aware of the Freiburg University dissertation filed a protest to the validity of the patent. The protest included testimony as to the date the dissertation was submitted and the procedure at the university library for indexing dissertations in the library catalog to make them accessible to the public. The examiner at the PTO rejected the reissue application, declaring the patent invalid, and the inventor appealed, first to the Board of Patent Appeals and Interferences and finally to the Court of Appeals for the Federal Circuit,[17] both of which agreed with the examiner. The inventor argued that the presence of a single cataloged thesis in a single university library did not constitute sufficient accessibility of the contents of the thesis to those interested in the field of the invention to amount to a "printed publication." The inventor also argued that the only evidence that had been presented to the examiner in the protest was the date on which the dissertation was submitted and the general procedure in the library for indexing dissertations, including the length of time it normally took to complete the indexing. The Court rejected both arguments, explaining that a single cataloged thesis was indeed sufficient to make the thesis accessible; that while evidence of a specific date of cataloging and shelving would be desirable, the "realities of routine business" indicate that a specific date is not strictly necessary; and that competent evidence of the general library practice, such as that presented, was sufficient to establish that the cataloging and shelving occurred before the critical date.

3.2.4 Grant Proposals

The submission of a grant proposal to a government agency is another common practice among university-based researchers, and the public's accessibility to the proposal that gives the proposal the status of a "printed publication" arises from the Freedom of Information Act[18] (**FOIA**). Similar to the thesis case of Section 3.2.3, the grant proposal in the case presented in this section was one submitted by a party independent of the inventor. Here again, however, the principles are the same as if the grant proposal had been submitted by the inventor.

---

[17] *In re* Leo M. Hall, 781 F.2d 897 (Fed. Cir. 1986).

[18] Title 5 of the U.S. Code, Section 552.

The invention in Patent No. 4,683,202[19] was the now well-known and widely used polymerase chain reaction (**PCR**), which is a laboratory procedure for generating millions of copies of a strand of DNA from a single copy (or from very few copies), the millions of copies being necessary for further study of the DNA, including determining its nucleic acid sequence. The process, known as "amplifying" DNA, had been developed at Cetus Corporation, the patent owner, and has become a powerful tool in DNA research. The validity of the patent was challenged by E.I. Du Pont de Nemours and Co. on the basis of a grant proposal submitted by Dr. Gobind Khorana at the Massachusetts Institute of Technology to the National Science Foundation (**NSF**) before the critical date of the Cetus patent. Du Pont claimed that the grant proposal was prior art against the Cetus patent since the proposal amounted to a printed publication. Cetus argued that the grant proposal, even though it was available to the public under the FOIA, was not prior art since it was not adequately stored and indexed by the NSF and thus was not sufficiently accessible to the interested segment of the public to constitute a printed publication, particularly since it had a vague title ("Chemical and Biological Studies of Nucleic Acids"), which did not by itself suggest the amplification of DNA.

Aside from the vagueness of the title, a question had been raised as to whether the information in the grant proposal was sufficiently detailed to disclose the invention actually claimed in the patent or simply close enough to raise questions of novelty or nonobviousness. The issue addressed by the Court however was the more fundamental question of whether the proposal was prior art as a printed publication preceding the patent's critical date, since if it were not prior art, Cetus would have no need to distinguish its invention over the descriptions in the proposal. The Court decided that the proposal was indeed a printed publication[20] preceding the critical date and hence prior art. Using the same standard as the Court in the case of the thesis in the university library, the Court in the Cetus case held that given the advanced technology in information storage and retrieval, the governing standard in determining whether a document was a "printed publication" was its public accessibility. The grant proposal had been indexed by title, author, institution, and grant number in the NSF's published indices of grants and awards and was available upon request from the NSF under the FOIA, all before the critical date, and therefore sufficiently accessible to researchers in the field to qualify as a "printed publication." The court further noted that Dr. Khorana was well known as a preeminent researcher in the field of DNA at the time of the proposal, which made it particularly likely that a researcher in the field would find the grant proposal by exercising reasonable diligence.

Since the accessibility of the grant proposal was through the FOIA, the date of the grant proposal as a "printed publication" was the date on which the NSF allowed the public access to the proposal, and this is true of all grant proposals and the funding agencies to which they are submitted. In the case of the NSF, public access is allowed only upon the agency's decision to award the grant. Other agencies that

---

[19] "Process for Amplifying Nucleic Acid Sequences," United States Patent No. 4,683,202; Mullis, inventor; Cetus Corporation, assignee; issued July 28, 1987.

[20] *E.I. Du Pont de Nemours & Co. v. Cetus Corporation*, 19 USPQ 2d 1174 (1990).

adopt the same policy include the Fund for the Improvement of Postsecondary Education (**FIPSE**) and the National Institutes of Health (**NIH**). Certain agencies, for example, the Environmental Protection Agency (**EPA**), make grant proposals available regardless of whether they are awarded, and thus proposals that are rejected by these agencies are prior art if made before the critical date. Each agency has its own procedures, which themselves are available upon request, including procedures for preventing the agency from publicly releasing the contents of the proposal.

3.3 OBSERVATIONS

The rules in the cases cited herein for both the on-sale bar and the publication bar may appear to be straightforward, but applying them to different sets of facts shows the limited control that one has over the result. The rule in the August Technology case is that if an offer is made before the invention is fully conceived and completion of the conception occurs later, the 1-year time period will start with the completion if the offer has not expired or been withdrawn. Completion may result in only a rudimentary conception, however, and the point at which this occurs can be arguable since completion of a conception turns on the functionality of the invention as conceived, and if additional development occurs before achieving a final design that is fully satisfactory, functionality at the conception stage can be highly debatable. The rule in the Atlanta Attachment case is that problems with certain features of an invention at the time of a sale or offer do not prevent the invention from being ready for patenting unless those features are ultimately reflected in the claims. Features that are not explicitly claimed may be implicit however or may affect the operability of those features that are explicitly claimed, in either case obscuring the distinction between claimed and unclaimed features. As for the experimental use cases, a factor that is commonly cited when testing is done by the buyer is the degree of supervision and control of the testing by the inventor or the inventor's agent, which is susceptible to argument depending on whose interest is at stake. In the publication cases, public access is a matter of degree, such as the degree to which an archived item is cataloged and indexed and the searchability of the item, and is also susceptible to difficulties in determining when a particular item is actually made accessible to the public.

Thus, while the lines of demarcation between the imposition and the avoidance of either bar may be easily stated, actual fact situations are often very close to those lines, and slight deviations can make the difference between falling on one side of the line and falling on the other. Avoiding any such activities before the critical date, that is, filing one's patent application within the year following any such activity, will provide the best assurance of steering clear of either bar.

Chapter 4

Canceling Prior Art and Other Benefits of Record Keeping

As the preceding chapters demonstrate, the conversion under the AIA from the first-to-invent standard to the first-to-file standard has fundamentally changed both the manner in which priority disputes are resolved and the definition of prior art. Where patent applicants have previously been able to prevail over competing patent applicants as well as eliminate certain prior art by producing laboratory notebook pages, internal memoranda, research reports, and other documentation of an early invention date, the AIA has shifted the applicant's emphasis from this type of documentation to the effective filing date of the patent application, and corporate strategies are being retooled accordingly. What value then does record keeping still have? It has considerable value, because the AIA has its own provisions that allow or require inventors to produce information that is best found in well-kept records, both for purposes of competing with others for patents on the same invention and in addressing prior art. Well-kept records are useful in other areas as well, including those that have not been changed by the AIA. This chapter begins by addressing situations where records showing that information was derived from an inventor, with or without the inventor's authorization, can work to the inventor's advantage, and continues with situations where records showing that an invention was a collaborative effort can also be of benefit.

4.1 DERIVATION PROCEEDINGS

A **derivation proceeding**, newly introduced by the AIA, is a proceeding for determining whether an invention claimed in a patent application was derived from someone other than the inventor(s) named in the application rather than being an original invention of the named inventor(s). The proceeding is conducted within the Patent and Trademark Office (PTO), which, if it is determined that the invention was

First to File: Patents for Today's Scientist and Engineer, First Edition. M. Henry Heines.
© 2014 the American Institute of Chemical Engineers, Inc. Published 2014 by John Wiley & Sons, Inc.

indeed derived, will correct the naming of the inventor(s) to include or substitute, as appropriate, the individual(s) from whom the invention was derived. A person who claims to have been the source of the invention and wishes to petition the PTO to institute the proceeding must also have a patent application pending that claims the same, or substantially the same, invention and must state in the petition that the application being complained of was filed without the petitioner's authorization. Derivation proceedings apply only to patent applications with effective filing dates falling on or after March 16, 2013.

Since derivation proceedings are contests between two applications claiming the same invention and are resolved in the PTO, derivation proceedings have a certain common ground with the **interferences** that are part of the pre-AIA system (interferences still being applicable to patent applications that have effective filing dates prior to March 16, 2013). The AIA itself suggests a parallel between the two proceedings by eliminating references to interferences in the patent statute and in at least one location by replacing the word "interference" with "derivation proceeding." Interferences are priority disputes, however, while derivation proceedings are inventorship disputes. Interferences thus determine which of two or more parties separately applying for patents on the same (or substantially the same) invention was the first to invent the invention and is therefore entitled to a patent on the invention. This can include derivation-type conflicts, that is, those in which one party claims that the other party had derived the invention from the first party, but the scope of the matters to be resolved in an interference and their manner of resolution reflect the first-to-invent rule and therefore extend beyond questions of derivation. An interference can thus arise between patent applications that had each been filed innocently by separate parties, neither of which had knowledge of the other's activities, and the parties can both have been actual inventors independently conceiving the invention and reducing it to practice. The losing party in an interference may simply have had a late start on coming up with the idea or on reducing it to practice, or if not actually reducing it to practice, then on diligently filing a patent application. In certain circumstances, therefore, the party with the later filing date can be the winning party in an interference. The party with the later filing date can also be the winning party in a derivation proceeding, but if it is, it will have its name placed on the opposing party's application rather than receiving a patent on its own application. This gives the winning party an earlier filing date than that of its own application. Note that the **petitioner** in a derivation proceeding, that is, the party who maintains that the opposing party (known as the "**respondent**") merely derived the invention from information that it received from the petitioner, will always be the later of the two parties to file a patent application (i.e., the party with the later effective filing date), since if it were the earlier, the first-to-file rule would directly entitle it to the patent. A derivation proceeding is thus a means of opposing the application of the first-to-file rule.

Evidentiary showings, and specifically record keeping, are integral to derivation proceedings as they are to interferences and are critical to both petitioner and respondent. The petitioner thus needs to provide documents that show both:

a. That the invention claimed by the respondent was derived from the petitioner (or from an inventor named in the petitioner's application)

b. That the respondent filed its patent application without authorization from the petitioner

The necessary documentation therefore includes records showing communication from the petitioner to the respondent, including the date and the means of the communication, the individual to whom the communication was directed, and the information communicated, or any other factual evidence establishing a connection between the petitioner as source and the respondent as recipient. The closer the information is in scope to the invention claimed in the respondent's application and the closer the information is to a completed conception of the invention, with or without verifying test data, the greater the likelihood that the petition will be granted. Records showing these facts must be verified, that is, accompanied by a statement by the petitioner of their accuracy and truthfulness, and corroborated, that is, their existence and date confirmed either by a witness other than the person who created the record or by other means of independently establishing the authenticity of the record, such as proof of an electronic communication or a return communication from the respondent acknowledging the communication. The respondent has the opportunity to rebut the petitioner's claim of derivation with evidence showing that the respondent conceived the invention independently of the petitioner. This can include, for example, origins of the idea that are independent of the petitioner, development or completion of the concept other than through the petitioner, differences between the scope and content of the information received from the petitioner and the invention, and dates of activity relating to the respondent's invention that predate any information received from the petitioner, including origination, exploration, testing, and recordation. The respondent's evidence will likewise require verification and corroboration.

A petition to institute a proceeding must be filed by a date that is 1 year or less from the date of the first publication of any claim to the invention that the petitioner believes to have been derived, and the petitioner must send the respondent full copies of the petition and supporting documentation within the same 1-year period. The appearance of the claim in published form that marks the start of the 1-year period can be either in a published patent application or a granted patent, the application or patent having been filed either by the petitioner or by the respondent. The simplest scenario will be one in which the respondent has filed a patent application and the petitioner files a patent application at a later date either before or after discovering the respondent's application. Since both applications are published 18 months after their filing dates, the respondent's application is published first since it has the earlier filing date, and the 1-year time limitation starts with that publication. Scenarios in which the claim first appears in a published version of the petitioner's application rather than the respondent's application can arise when the petitioner's application claims a filing date benefit from an earlier (e.g., parent) application that does not itself contain the claim. The publication then occurs 18 months from the petitioner's earlier application despite the fact that the earlier application does not contain the claim of interest. In either case, the triggering publication can also be in a patent application filed in a jurisdiction other than the PTO, such as through the Patent Cooperation Treaty or a non-U.S. patent office such as the European Patent Office.

Derivation proceedings can also be conducted in federal court rather than in the PTO, although different rules and procedures will apply, and can be between two or more patents or between one or more patents and one or more applications. In all cases, the proceedings and outcome will be determined by the same type of evidence as those conducted within the PTO.

4.2 DISQUALIFYING REFERENCE MATERIALS AS PRIOR ART

Patents and pending patent applications frequently disclose information beyond what they recite in any of their claims, and these documents are commonly cited as prior art for this information in the same manner as published materials in general. The variety of materials that can be cited for these disclosures, including those originating or published both in the United States and outside the United States, is listed in Chapter 2, and as prior art, they raise obstacles to an inventor's ability to obtain patent coverage even though they are not an attempt to compete with the inventor for patent coverage on that information. According to the AIA, however, these documents and materials are *not* prior art if their authors "obtained the subject matter disclosed directly or indirectly from the inventor or a **joint inventor**." The AIA places this wording in multiple locations within the patent statute, including:

1. As exceptions to printed publications (including patents and published patent applications anywhere in the world and any other published material) that are prior art under Section 102(a)(1) with publication dates that precede the inventor's effective filing date
2. As exceptions to U.S. patents and published U.S. patent applications that are prior art under Section 102(a)(2) regardless of when they were granted or published

These exceptions bear a surface similarity to the fact situations that serve as the basis for derivation proceedings—compare the wording "obtained the subject matter ... directly or indirectly from the inventor" (the wording of the exceptions) with "derived from an inventor named in the petitioner's application" (the wording describing the reason for a derivation proceeding). Unlike derivation proceedings, however, these exceptions are not a contest between competing claims to the same invention but rather a determination as to whether the subject matter of a claim has been previously disclosed and therefore lacks novelty. Also, authorization or the lack of authorization for their publication from the inventor is irrelevant to whether a document or material falls within either exception, and no petition or special proceeding is required for invoking either exception.

In the typical examination of a patent application at the PTO, an examiner will conduct a search and select materials from those produced by the search for presentation to the applicant as prior art, generally without knowing the source of information in the materials other than what the materials themselves indicate. In most cases, therefore, the examiner will be unaware that a given material falls within one of the exceptions to prior art. The applicant can then respond by showing that an exception does indeed apply and can do so by submitting a verified statement showing that the inventor was the source of the information. Such a statement will require supporting documentation such as a record of a communication to the author of the cited material, including the date and the means of the communication and the actual information communicated. Here again, the closer the communicated information is to the disclosed subject matter, the more persuasive the communication will be, and the date of the communication must be consistent with the assertion that the published information was obtained as a result of the communication. Corroboration (witnessing or other independent authentication) of the statement is not required, but enhances its credibility when included.

4.3 RECORDS SHOWING COLLABORATION

Patents that list two or more individuals as **coinventors**, also referred to as "joint inventors," typically originate in research institutions or corporate environments where researchers collaborate and assign their rights to a common entity, usually their employer. When coinventors assign to different entities or retain their rights individually, however, each retains certain rights that may be to their benefit but may also limit the ability of the others to enjoy certain benefits that they would have if they were sole inventors. The sharing and limitations of rights between coinventors can raise conflicts when different inventors have different contractual obligations, such as those in typical employment agreements requiring them to assign their patent rights to their individual employers. This is illustrated by a case history that shows how evidence of a collaborative effort and hence joint inventorship might have prevented a manufacturer from being held liable for patent infringement.

This case[1] involved a patent on expanded polytetrafluoroethylene (ePTFE), which is a polymer that has a weblike microstructure consisting of nodes interconnected by fibrils. This microstructure causes fabrics made from this polymer to be both waterproof and breathable, and a well-known commercial product made from this polymer is Gore-Tex, developed by W.L. Gore & Associates, Inc., a fabric used for rainwear. The polymer has other uses, however, one of which is as a material for vascular prostheses (grafts for blood vessels) since the weblike microstructure allows the polymer to support cellular ingrowth from adjacent vascular tissue to anchor the prosthesis in a vein or artery. The use of ePTFE for vascular prostheses is the subject of Patent No. 6,436,135,[2] owned by Bard Peripheral Vascular, Inc. The patent states that for the cellular ingrowth to be uniform and controlled and for the web to promote the formation of a thin, viable neointima (the innermost layer of a vein or artery), qualities that are essential for a successful prosthesis, the microstructure must meet certain requirements,

---

[1] *Bard Peripheral Vascular, Inc., et al., v. W.L. Gore & Associates., Inc.*, 670 F.3d 1171 (Fed. Cir. 2012).

[2] "Prosthetic Vascular Graft," United States Patent No. 6,436,135; Goldfarb, inventor; issued August 20, 2002.

specifically an average fibril length within a specified range of 6–80 microns. A researcher at Gore was also exploring the use of ePTFE for vascular grafts and came to the same conclusion as that stated in the Bard patent, that is, that the 6–80 micron range was critical to the success of the graft. As part of his investigation, the Gore researcher provided tubes of ePTFE to researchers outside the company for evaluation, one of whom was the individual who ultimately applied for the patent and assigned it to Bard. Aside from the Bard researcher receiving the tubes from the Gore researchers, the two worked independently. The Gore researcher also applied for a patent citing the critical range, and since the two patent applications were both pre-AIA, an interference ensued, with the Bard researcher prevailing over the Gore researcher by showing a reduction to practice of the invention earlier than the Gore researcher. By the time that the interference was resolved, Gore had been manufacturing vascular grafts of ePTFE with fibril lengths within the critical range, and Bard sued Gore for infringement.

One of the strategies that Gore adopted for responding to the lawsuit was to claim that its own researcher was a coinventor of the patent, in view of the fact that the tubes that the Bard inventor had used in his studies had been supplied to him by the Gore researcher. When a patent lists coinventors, patent law states that each coinventor can assign his or her rights independently of the other coinventor(s), each can grant licenses under the patent without sharing the royalties with the others, and each can practice the invention without seeking permission from, or paying royalties to, the other(s). If Gore could have established that its own researcher was a coinventor on the Bard patent, Gore would have had full use of the invention in the patent independently of Bard. Otherwise, Bard would be entitled to damages from Gore in the hundreds of millions of dollars for patent infringement.

The testimony at trial showed that the microstructures of the tubes that Gore supplied to the Bard researcher varied, some having average fibril lengths within the critical range and others outside the range. The testimony also indicated that the Gore researcher did not provide the Bard researcher with any information regarding the fibril lengths, much less how different lengths might affect the success of the tubes as graft material. Patent law states that one of the qualifications of joint inventorship is collaboration or concerted effort between the inventors, and another is that every coinventor must contribute in some significant manner to the conception or reduction to practice of the invention as claimed. Since the fibril length was critical to the success of the ePTFE in a vascular graft and a claimed feature of the invention and since neither researcher conveyed his findings to the other, the court found that collaboration and concerted effort were lacking and that by failing to provide the Bard researcher with his own knowledge about the critical length, the Gore researcher had not made a significant contribution to the invention in the Bard patent.

Companies typically discourage their employees from releasing proprietary information outside the company since releasing such information compromises the company's intellectual property rights. If information is exchanged between companies and the companies wish to preserve intellectual property in the form of a patent that includes contributions from both companies, the respective rights and obligations of each company are typically agreed upon in advance. Joint ownership between companies without restrictions is generally avoided since it can lessen one company's competitive position relative to the other. In this case, joint ownership would have lessened Bard's competitive position relative to Gore since Gore would have been free to manufacture and sell vascular prostheses made from ePTFE with the fibril lengths claimed in the patent, without owing anything to Bard. This illustrates the need for clearly establishing the mutual rights and obligations of parties before sharing any information or materials and that in certain circumstances, there is an advantage in fully revealing one's findings when utilizing outside entities for testing or similar purposes so that the origination of an invention is made clear.

4.4   RECORDS OF PUBLIC DISCLOSURES AND COMMERCIAL USES

Public disclosure of an invention before filing a patent application can be valuable in some cases in securing the discloser's patent rights, and a commercial use, with or without filing a patent application, can be useful in defending the discloser from an accusation of infringement of another's patent rights.

As noted in Chapter 2, public disclosures by the inventor or by someone who obtained the disclosed information from the inventor and whose disclosure dates are no more than 1 year before the effective filing date of a patent application are not prior art themselves but can be used to disqualify other public disclosures as prior art if they predate the other public disclosures. Thus, information contained in a technical paper or symposium presentation, or in an archived doctoral thesis, for example, that are authored neither by the inventor nor by someone having any connection with the inventor can be eliminated as prior art by showing that an earlier public disclosure of the same information was made by the inventor or by someone receiving the information from the inventor. The earlier disclosure may be in an entirely different form, medium, or venue than the disclosure being disqualified, but will nevertheless have its disqualifying effect provided the disclosure is public and the patent application is filed within the following year.

Compared to internal records and records of communications to individuals outside one's company, public disclosures of an invention are generally easier to maintain and produce when needed as evidence, in view of their public nature. To rely on an earlier public disclosure, however, the inventor must produce not only a copy of the disclosure but also evidence (or at least a verified statement) of how the disclosure was made (i.e., in what form it was presented), to whom it was made or how it was made accessible to the interested public, and the date on which it was made. For disclosures made by others who obtained the information from the inventor, evidence will also be needed showing how and when the information was conveyed by, or otherwise obtained from, the inventor.

The inverse of overcoming prior art by showing one's own prior public disclosure (or a prior public disclosure by one's recipient or agent) within 1 year of one's own filing date is an accused infringer overcoming an assertion of infringement by showing that the infringer itself had commercially used the invention more than 1 year prior to the patent's filing date or the patentee's prefiling public disclosure. This showing must be of a commercial use rather than a public disclosure, and it must show that the use occurred in the United States and more than 1 year prior to the patentee's filing date rather than less. While a commercial use this early may also invalidate the claims of the patent that cover it, this "**prior use defense**" is designed for directly responding to an accusation of infringement rather than requiring the infringer to argue that the claims are invalid, which is often a more difficult case to make. The defense applies only to the use that actually occurred and only to the person who performed or directed the use. In addition to commercial uses, the defense also applies to noncommercial uses by nonprofit entities such as hospitals and universities so long as the public is the intended beneficiary of the use and any continued use by the entity itself is noncommercial. The defense also applies to uses of drugs for purposes of premarketing regulatory review. The defense applies to any patent issued on or after September 16, 2011 (as opposed to the first-to-file provisions that became effective on or after March 16, 2013).

Prior to the AIA, the prior use defense applied only to patents on business methods, and the AIA expanded its applicability to any process, machine, manufacture, or composition of matter. The AIA introduced a limitation however in cases where the inventor in the patent had made a prefiling public disclosure less than a year before the filing date, the limitation being in the dates on which the infringer's prior use must occur to qualify as a defense. While previously the use could occur up to 1 year before the effective filing date of the patent whether or not there had been a prefiling public disclosure by the inventor, the AIA states that if such a disclosure occurred, the use must occur prior to 1 year before the disclosure.

Commercial uses may be more difficult to record, organize, and maintain than public disclosures, since the purposes of any documentation of a commercial use will typically be even further removed from its use as a defense to infringement. For this reason, special care may be necessary to preserve the factual evidence of a commercial use in a form that supports the defense, such as a clear description or identification of what was actually "used" rather than simply the dates of the use and other numbers (such as the number of runs performed, the volume sold, the revenue generated, etc.). In addition to showing the earliest dates of use, the record must show that the predominant purpose of the use was commercial rather than experimental, and useful information to support that can include the volume of the use, the customer or other beneficiary of the use, the revenue generated by the use including the user's profit, and the absence (or minimal amount) of mutual exchanges between user and beneficiary regarding further investigations or testing. If the use was a continuing use through the 1-year date, the records should show this as well, since the defense will not apply if the use was abandoned, even temporarily, before the 1-year date.

4.5 LABORATORY NOTEBOOKS

An entry in a laboratory notebook is typically made at the time the information in the entry is created or soon thereafter, without extra time spent in compiling, organizing, and analyzing the information or in deciding upon recommendations for further action, as often included in more formalized reports to management. For this reason, the laboratory notebook is one of the most useful forms of record keeping, particularly for derivation proceedings and submissions by an inventor to disqualify prior art. A laboratory notebook can show the individual in whom an idea originated and when, and in joint efforts, who contributed what. The keystone of the act of invention is generally the conception itself, and the individual who merely instructs the other(s) as to the state of the art by providing information that is already known, or who performs routine testing, synthesis, or prototype construction at the request of the other(s), is neither an inventor nor a coinventor.

To be effective as evidence, each notebook page must bear the signature of the person making the entries and the date on which the entries were made. The signature of a witness can also be included as corroboration that the page existed in completed form as of the witnessing date. Many notebook pages contain the preprinted words "witnessed and understood by" or equivalent wording, preceding the witness's signature blank to indicate that the witness signing the page read the entries on the page before signing and confirms their contents rather than simply their existence. Entries made on a notebook page should be legible and explanatory, including a narrative of what was conceived or actually done, preferably with explanations of any acronyms, words, or symbols that might not be understood by others. It is important to remember that while notebooks are kept primarily for the benefit of the researcher, they will also be read by people with no connection to the researcher and less of an understanding of the relevant technology when they are presented as evidence before the PTO or in court.

Electronic records can serve the same purpose as handwritten records and, in addition to eliminating the legibility problem, are often more detailed, since most people can enter information more quickly in electronic form and thereby enter the information in greater amounts. Unfortunately, electronic records tend to lack some of the features that are routinely included in handwritten notebooks, such as identification of the person creating the record, the date on which the record was created, and the contents of the record when it was created (as distinct from modifications or additions entered subsequently).

To be useful for the purposes set forth in this chapter, an electronic record must not only identify the person making the record and the date of the record but must also be reproducible in human-readable form long after the record's creation. Up to 10 years may pass between creation of the record and the need for its use by an inventor attempting to get a pending patent application allowed by the PTO, and as many as 30 years may elapse if the record is used as evidence in a lawsuit. Since software changes are likely to occur during these periods, software for reading the records in obsolete formats must be maintained or be otherwise available.

Electronic records should be archived in a readily accessible and identifiable form and maintained in a manner that prevents them from being modified, since modification can compromise one's ability to establish the date of the original entry as well as the contents. Employers can improve the chances that the records are being handled in this manner by setting policies and incorporating mechanisms for archiving electronic records. A record custodian can be assigned the task of managing the records in an ongoing manner.

Chapter 5

Inventing in an Employment Environment: The AIA's New Recognition of Employer Interests and Project Management

It has always been a goal of intellectual property law to encourage creativity, to promote the exchange of new ideas, and to advance the state of the art by allowing individuals who do so to reap the rewards of their efforts. As a result, the legal system has traditionally favored the inventor when dealing with conflicts between the inventor and a business entity, particularly the inventor's employer that claims an interest in the invention. Employers do have legitimate interests, of course, and as business and commerce continually expand, there has been an increasing recognition that laws and regulations that are centered on the inventor at times run contrary to the practicalities of the employment environment by failing to recognize the contribution that the environment itself can make to the inventing process. In some cases, this has placed employers at a disadvantage. The AIA recognizes this and introduces significant changes in the status of employers in the process of obtaining patents. These changes appear in two principal areas—defining what is and is not prior art and allowing the employer to substitute for the inventor in the procedure for filing a patent application.

First to File: Patents for Today's Scientist and Engineer, First Edition. M. Henry Heines.
© 2014 the American Institute of Chemical Engineers, Inc. Published 2014 by John Wiley & Sons, Inc.

5.1   PROJECT MANAGEMENT AND THE NEW DEFINITION OF PRIOR ART

As noted in Chapter 2, one subclass of prior art that can be cited against a patent application to challenge the novelty (and nonobviousness) of the invention is earlier-filed patents and published patent applications that list at least one inventor other than the inventor of the challenged application. The earlier filing date of the reference patent or published application and the fact that a different inventor is named in the reference thus qualify the reference as prior art, regardless of whether the issue date or publication date of the reference predates or postdates the filing date of the challenged application or whether the inventor in the challenged application is also listed as an inventor on the reference. If all of the inventors between the challenged application and the reference are employees of a common employer, the reference can be difficult prior art to overcome, since the two inventions often arise within a common project or relate to a common product of the employer. When coemployees work under the same roof, there is a high likelihood that they will share and compare results, and such an exchange of information will not violate the employer's confidentiality policies since it does not involve disclosures outside the company. Indeed, the exchange of information among coemployees is typically encouraged, if not required, by the employer, by including project reports to managers and periodic group meetings with managers and coworkers among the business routine. Presentations of one employee's work to other employees working on different aspects of a common project allow all to benefit from their coworkers' findings and to offer contributions of their own to the presenter for the general benefit of the project. Project managers will often use the information reported in these presentations to make managerial decisions, including shifting or refocusing work on the project or increasing or otherwise adjusting the project's funding or staffing. In-house reporting can also boost employee morale by giving the employee the satisfaction of having his or her work recognized and appreciated. With such internal exchanges of information, it is not unlikely that one employee's patent application will make mention of or reference to the work of another employee that has not yet been considered for patent protection but may be at a future date and consequently may find itself rejected over the statement in the earlier application. The same can also be true outside the common-employer situation, such as in a joint research arrangement.

In recognition of this reality, the AIA specifically recognizes earlier-filed patents and published patent applications that name among their inventors an individual who is not named in a later-filed application and treats these earlier-filed documents as exceptions to prior art against the later-filed application, as long as the earlier-filed and later-filed are either commonly owned or subject to an obligation of assignment to the same entity. For this exception to apply, the common ownership or obligation must have been in effect at the effective filing date of the later-filed invention, and the reference patent or published application must have an issue or publication date, respectively, that was not more than 1 year prior to the effective filing date of the later-filed invention.

A caveat to this new exception to prior art is that it does not permit the employer to claim the same invention in two separate patent applications. Doing so would constitute "**double patenting**," which is in violation of another section of the statute that limits the inventors and their employer to one patent per invention. Even when the invention claimed in the later-filed application is not exactly the same as the invention in the earlier, however, the two are often very close. This occurs, for example, when the later application is filed to refine or somehow extend the invention in the earlier-filed application in a relatively minor way that might be considered "obvious," such as expressing the invention differently but not sufficiently so to be patentable on its own over the earlier-filed invention (the nonobviousness requirement is discussed in Chapter 7). This is called "**obviousness-type double patenting**" and is likewise prohibited. The concern in both double patenting and obviousness-type double patenting is that if a common owner were to obtain patents on two applications for what is essentially a single invention, and the normal expiration date of one of the patents is later than that of the other, the owner would effectively receive a total term of patent coverage for a single invention that would exceed the term set by the patent statute. This can be avoided by making sure that the two inventions differ in some way, even a relatively trivial way, and limiting the expiration date of patent that might issue on the later-filed application to coincide with the expiration date of the patent that issued (or might issue) on the earlier-filed application. This limitation of the term is achieved by filing a "**terminal disclaimer**," that is, a statement that disclaims the terminal portion of the patent that would be the later to expire, that is, it cuts short that patent's term of enforcement, to cause both patents to expire on the same date.

The prohibition against both double patenting and obviousness-type double patenting and the means to avoid or overcome them are not new to patent law with the AIA but instead remain unchanged from pre-AIA law. The AIA merely eliminates the prior art status of the earlier patent or application. This is a significant change, however, since prior art that is not otherwise disqualified must be distinguished over by argument or explanation. The AIA thus eliminates the need for the patent owner to make statements on the PTO record (which is accessible to the public) that might belittle or limit any information contained in an earlier application or patent in order to obtain grant of a patent on a later application.

5.2   ALLOWING THE EMPLOYER TO STAND IN FOR THE INVENTOR

It is clearly to the advantage of the patent application and its owner if the inventor is closely involved in the preparation of the application. Inventor involvement in the preparation of the application assures accuracy in both the **specification** (the text of the application preceding the claims) and the claims, including an accurate statement of the invention both in its own terms and in relation to the state of the art, since the inventor will likely be the one most familiar with the pertinent prior art, including the sources that the inventor used in developing the idea. The inventor's name will appear prominently on the cover sheets of both the application when published and the patent when issued, and for this reason, the contents of both documents will be attributed to the inventor even when much (or even all) of the content has been drafted by a patent attorney. Attribution to the inventor is particularly likely since the specification is intended to teach the invention to, and is thereby expressed in terms directed to, those "skilled in the art," that is, persons with education in and an up-to-date familiarity with the field of technology of the invention. It is natural to assume that the inventor is among those skilled in the art and therefore an appropriate person to teach the invention to others.

It is also to the advantage of the employed inventor to cooperate with the employer by executing not only the inventor's declaration that accompanies the patent application but also an assignment of the invention to the employer, as employees are generally obligated to do by the terms of their employment. The obligation typically continues after the employee has left the company, in which case the advantage is less evident, particularly when the individual's new employer is a competitor of the former employer and in situations where the departure was not amicable. It may go against one's nature to maintain favorable relations with a former employer when the departure was the result of a disagreement or when the employee was forced to leave. Nevertheless, the individual's reputation in the industry can be hurt by continuing to be hostile to a former employer, and future opportunities may later arise with the same employer or with coworkers or coinventors, which opportunities may be unnecessarily jeopardized by a refusal to cooperate in meeting one's obligation to the former employer.

Situations will nevertheless occur where an assignment from an inventor employee (or former employee) cannot be obtained. These include the death of the inventor, the legal incapacity of the inventor,[1] and the outright refusal of the inventor to assign despite an obligation to assign. The PTO has long recognized the difficulties associated with deceased inventors and inventors who are under a legal incapacity and has accepted substitute statements and signatures from heirs, executors, guardians, conservators, and other legally recognized representatives of these inventors. For inventors who are under an obligation to assign to their employers but refuse to do so, the AIA adds the employer to those who can step in on the behalf of the inventor and also confers ownership of the patent directly on the employer despite the inventor's refusal to do so. While an employer claiming its employee's obligation to assign could always seek to enforce the obligation in a court of law, the AIA's provision for issuing a patent to its "obligated assignee" (the inventor's employer) eliminates the need for a legal confrontation with the employee and simplifies the procedure considerably.

---

[1] A person who does not comprehend the nature and consequences of a contract that the person is being asked to sign (or, in this case, the declaration that inventors are required to sign when a patent application is submitted to the PTO and the assignment of the invention to the inventor's employer pursuant to the employment agreement) is considered to be under a legal incapacity to sign. Such a person can be declared incompetent or legally incapable by a court of law that will appoint a guardian or conservator to sign the assignment document in the person's place.

5.3 WHAT CONSTITUTES AN OBLIGATION TO ASSIGN?

The most common form of the obligation is an express agreement, typically found as a provision in a general employment agreement or in a consulting contract, joint research or development agreement, or other documents recognizing the possibility that patentable discoveries may arise during the course of the relationship between the parties and setting in advance the ownership of any intellectual property rights resulting from the discoveries. Employees are commonly asked to sign a general employment agreement that contains such a provision on the first day of their new job. Companies neglecting to do this and later coming to appreciate the value of such a provision often require their existing employees to sign a supplemental agreement, and either such agreement is enforceable. When no agreement has been signed by the employee, the obligation can still exist in the form of a company policy issued by the employer and circulated to all employees. To rely on such a policy statement, the employer may be asked to establish that an unwilling employee was actually aware of the policy statement at the time the invention in question was conceived and (at least tacitly) acquiesced to it. Unless challenged by the employee, however, the PTO requires only a "clear and conspicuous statement" by the obligated assignee's patent attorney of the existence of an obligation to assign and does not require a copy of the actual agreement.

An agreement to assign inventions can be a precondition to the acceptance of a job or to the continuation of one's employment, and courts in both cases will generally recognize the agreement as legitimate, even when the employee has little or no choice. The courts' view is that the employee's salary is intended to cover, and is reasonable compensation for, the innovations that the employee produces in connection with his or her employment and that the typical employee is not without other options, since the employee can seek other employment in the same field using the same skills or in many cases draw an income through self-employment using those skills.

An illustration of an employee's violation of an obligation to assign is the case of *Cubic Corporation v. Marty.*[2] The invention in this case was a training system for military pilots whose aircraft is being targeted by radar homing missiles from the ground (surface-to-air missiles) or from enemy aircraft. Before the invention, training systems utilized radar generated by signal-emitting stations on the ground that mimicked the radar used by the homing missiles and thereby simulated an attack by one of these missiles. A receiver on the aircraft detected the mimicking radar in the same way that it would when the aircraft was under attack and warned the pilot to take appropriate evasive action. These systems presented problems, however, including interference of the mimicking signals with television and radio communications in the area in addition to the high cost of the equipment needed to produce the signals. The invention involved the use of a threat simulator on the aircraft itself, which takes

---

[2] *Cubic Corporation v. William B. Marty, Jr.*, 185 Cal. App. 3d 438 (California Court of Appeals, 1986).

coded digital information and converts it to the same type of warning that the pilot received with the previous system. The coded digital information is transmitted from the ground but eliminates the need for the mimicking radar and the resulting interference and cost.

The inventor's employer, Cubic Corporation, had a system in place for air combat training for pilots but not one that used electronic simulation. The inventor (Marty) passed his idea to Cubic, who funded an internal project to develop the idea, made Marty the program manager, gave him the services of another employee to help design the circuitry, and gave Marty a raise that was higher than the average raise that the company was giving its employees at that time, plus a bonus. Marty had already signed an employment agreement with Cubic that stated that he "assigns and hereby agrees to assign … all ideas, processes, inventions, improvements, developments and discoveries coming within the scope of Company's business or related to Company's products …." Despite the agreement, Marty hired his own patent attorney and through his attorney obtained Patent No. 4,176,468 on the new system.[3] The attorney then informed Cubic of the patent and proposed a license to Cubic. When Cubic responded that the patent was by rights already theirs and offered to reimburse Marty for his expenses, Marty refused and was ultimately terminated for doing so. Cubic then sued Marty for breach of contract (his employment agreement) and for a declaratory judgment that the patent belonged to Cubic.

Marty responded to the lawsuit with several arguments, including that the invention was not within the scope of his employment since he was not employed as a design engineer, that the employment agreement was an adhesion contract (i.e., one that has been drafted by one party with no opportunity for the other party to negotiate its terms or to look elsewhere for a more favorable contract), and that the compensation he received for the invention was inadequate considering the value of the invention. The court ruled against him, observing that adhesion contracts are still enforceable if they are not "unconscionable." In this case, Marty had been promoted from "electronics engineer" to "senior engineer" while employed at the company and evidence had been presented that showed that senior engineers' duties included design work. These considerations, plus the high raise and bonus, indicated that the agreement, or at least the manner in which Cubic honored the agreement, was not unconscionable and that Marty had been adequately compensated. The court emphasized that "the Labor Code provisions were not intended to award an employee who presents an invention to an employer, represents the invention is for the employer's benefit, actively seeks and obtains company funding to refine his invention, uses company time and funding to develop his invention while all the time secretly intending to take out a patent on the invention for himself."

Limitations on what the employer can require are well established by statute in several states, and blanket or overbroad obligations imposed by an employer are considered overreaching in all states by "common law," that is, law that is generally recognized by the legal system regardless of whether it is codified in a statute.

---

[3] "Cockpit Display Simulator for Electronic Countermeasure Training," United States Patent No. 4,176,468; Marty, Jr., inventor; issued December 4, 1979.

Obligations to assign that are unlimited in time, subject matter, or both are considered to be overreaching, and language that is ambiguous as to its scope of time or subject matter can likewise make an obligation unenforceable. An illustration of the type of language that can be considered ambiguous as to both time and subject matter is the case of *Mattel v. MGA Entertainment*.[4] The assignment provision in the Mattel's employment agreement read as follows:

> *I agree to communicate to the Company as promptly and fully as practicable all inventions (as defined below) conceived or reduced to practice by me (alone or jointly by others) at any time during my employment by the Company. I hereby assign to the Company ... all my right, title and interest in such inventions, and all my right, title and interest in any patents, copyrights, patent applications or copyright applications based thereon. ... [T]he term "inventions" includes, but is not limited to, all discoveries, improvements, processes, developments, designs, knowhow, data computer programs and formulae, whether patentable or unpatentable.*

The employee at the center of the lawsuit was an individual whose job at Mattel was to design fashions and hairstyles for high-end (collectible) dolls but separately conceived of a new line of dolls (with "attitude"). While still employed by Mattel, the employee pitched this new line to MGA Entertainment, Mattel's competitor, and MGA proceeded to develop the new line, calling it "Bratz" dolls. At its height, the Bratz line generated more than $1 billion a year in sales that cut deeply into Mattel's sales of its own line, the famous Barbie dolls. Mattel reacted by suing MGA and the employee for **copyright** infringement and breach of contract, based on the employee's employment agreement with Mattel containing the aforementioned provision. MGA responded by challenging the validity of the agreement.

MGA's challenge focused on the phrase "at any time during my employment by the Company" together with testimony by Mattel employees (other than the individual in question) as to how the company interpreted the phrase. One employee testified that it was "common knowledge that a lot of people were moonlighting and doing other work" and that this was understood to be acceptable so long as it was done on "their own time" and at "their own house," while another employee testified that her understanding was that "everything I did while I was working for Mattel belonged to Mattel." MGA's challenge also focused on the word "invention" in the agreement and whether this word was overly broad by covering ideas (such as dolls with "attitude"), despite the fact that ideas were not included in the types of inventions explicitly listed in the agreement (which said "but is not limited to" immediately before the list). A Mattel executive testified that it was common knowledge in the design industry that terms like "invention" and "design" included employee ideas. Considering all the testimony, the trial court found for Mattel, that is, that the agreement was valid and enforceable, and awarded Mattel $10 million in damages plus MGA's **trademark** portfolio on the Bratz line. MGA appealed, and the appeals court reversed the decision of the trial court.

---

[4] *Mattel, Inc. v. MGA Entertainment, Inc., et al.*, 616 F.3d 904 (9th Cir. 2010).

In reaching its decision, the appeals court observed that the expression "at any time during my employment" could either refer to the entire calendar period of the employment, including nights and weekends, or only to work hours. As for whether the agreement covered "ideas," the court observed that ideas are "ephemeral and often reflect bursts of inspiration that exist only in the mind," as opposed to designs, processes, computer programs, and formulae, which are "concrete," and that the omission of ideas from the list in the agreement indicated that ideas were not included in the agreement's definition of "inventions." For these reasons, both the expression "at any time during my employment" and the word "inventions" were held to be ambiguous.[5] While no patent was involved in the case, the agreement explicitly covered patents and inventions, and this, together with the court's emphasis on the scope of "inventions," indicates that the reasoning of the decision would have been applicable to patents as well.

It is interesting to compare the result in the Mattel case with the result in the pilot training case (*Cubic Corporation v. Marty*). Note that the possible inclusion of ideas within the meaning of "inventions" was deemed to make "inventions" ambiguous even though ideas were not explicitly included in the list and that if "inventions" did cover ideas, the subject matter of the obligation would have been considered overbroad. In the Cubic case, which was in the same jurisdiction as Mattel, the agreement explicitly listed "ideas" and yet was held to be enforceable. Because of the different natures of the inventions in the two cases and the differences between the types of industry that the employers were in, it appears that the subject of ideas did not come up in the Cubic case. On the other hand, the Mattel case was decided 24 years after the Cubic case, suggesting that in cases arising after Mattel, assignment obligations in employment agreements should not include "ideas" since this would cast doubt on the agreements' unenforceability.

Of the states that have enacted statutes limiting what an employer can require in terms of assignment obligations, seven states use the same language, which states that the employee cannot be obligated to assign an invention:

for which no equipment, supplies, facility or trade secret information of the employer was used and which was developed entirely on the employee's own time, and

(1) which does not relate
    (a) directly to the business of the employer or
    (b) to the employer's actual or demonstrably anticipated research or development,
or

(2) which does not result from any work performed by the employee for the employer.

---

[5] Once it determined that the terms were ambiguous, the appeals court sent the case back to the trial court to resolve the ambiguities. In this second round at the trial court, the case shifted from copyright to trade secret misappropriation, this time by Mattel rather than MGA, and the case went up once again on appeal, where all awards other than attorney fees were reversed. Nevertheless, the case illustrates how certain terms can be interpreted as overbroad (in this case, if the broader interpretations were ultimately deemed to be the ones to be applied).

States that have enacted statutes with this language are Minnesota, California, Delaware, Illinois, Kansas, North Carolina, and Washington. The corresponding statute in Utah is stated in broader terms:

> *(1)* *An employment agreement between an employee and his employer is not enforceable against the employee to the extent that the agreement requires the employee to assign or license, or to offer to assign or license, to the employer any right or intellectual property in or to an invention that is:*
> *(a)* *created by the employee entirely on his own time; and*
> *(b)* *not an employment invention.*

Whether this statute allows a Utah employer to assert a claim over an invention for which none of the employer's equipment, supplies, facilities, or **trade secret** information was used is not clear, nor is the scope of "employment invention." This statute may thus be broader in one aspect and narrower in another than those of the seven states listed earlier.

Nevada has taken a different approach by vesting the ownership directly in the employer as a default, that is, unless the employment agreement specifies otherwise. Nevada's statute reads as follows:

> *Except as otherwise provided by express written agreement, an employer is the sole owner of any patentable invention or trade secret developed by his or her employee during the course and scope of the employment that relates directly to work performed during the course and scope of the employment.*

Here as well, what is meant by "the course and scope of employment" may be open to interpretation. As noted earlier, however, employer rights and employee assignment obligations in all states, including Utah, Nevada, and the seven with the more explicit language, as well as those with no corresponding statutes, are subject to common law that holds that these rights and obligations cannot be unlimited in time or subject matter. The decision in the Mattel case indicates that actual practices or understandings within a company can impose interpretations on the terms in either statutes or employee agreements that can affect their enforceability.

5.4   IMPLYING AN OBLIGATION TO ASSIGN WHEN THERE IS NO EXPRESS AGREEMENT

The preceding section explains how "common law," which is judge-made law rather than statutory law but recognized nonetheless, imposes limits on how broadly an assignment provision in an employment contract can be interpreted. Common law also operates where there is no assignment provision or other written obligation to assign, by granting the employer ownership of, and hence imposing on employees an obligation to assign, inventions made by those who are "hired to invent." An employee who was hired to invent is one who was initially hired to solve a particular problem or to exercise his or her inventive faculties in a particular area of technology. Conversely, an employee hired for general service who invents on the side is not considered to have been hired to invent, and in the absence of an employment agreement that states otherwise, that employee's inventions remain the property of the employee.

Since the distinction between employees who are hired to invent and those who are not is not set out by the language of the typical employment agreement or a statute, the determination in particular cases is often less clear. The case of *Teets v. Chromalloy*[6] is an illustration of an employed inventor who was not under an express obligation to assign but was held to have been hired to invent a particular invention even though he conceived of the invention at home and after presenting the invention to his employer was instructed by the employer to focus his efforts during work hours on an alternate project.

The Teets case involved fan blades for commercial aircraft and particularly those used for the large thrust engines needed for carrying high loads. To get more fuel efficiency, metal fan blades were replaced with lightweight blades of a composite construction including a nonmetallic core and a cladding of a strong metal such as titanium over the core's leading edge. The cladding was to prevent the blade from fracturing in flight due to the impact of birds, freezing rain, and debris. The difficulty was in shaping the cladding and securely attaching it to the core, and Teets' employer, DRB Industries (a division of Chromalloy), was given the task by the General Electric Aircraft Company of producing a single-piece cladding and developing a way to attach it that would allow the cladding to retain its strength despite the stresses that it was exposed to during its attachment. DRB proposed manufacturing the cladding in sections and bonding them together over the core by either welding or diffusion bonding, but these methods were not succeeding. Teets, who was the chief engineer on the project, had an idea for a hot forming process that is shown in Figure 5.1A and B, which appears in Patent No. 5,694,683,[7] the patent that was ultimately involved in the lawsuit.

The process consisted of forming a blank (element 1 of Fig. 5.1A) of the cladding material with a trough (17) and relatively thick walls on either side of the trough, heating the blank to a temperature where the material was malleable, placing the heated blank in a mold cavity with movable walls (11 and 14), and pressing a mandrel (9) into the blank while the mold, blank, and mandrel were held at the elevated temperature. The mandrel had an outer contour that was the same as that of the nonmetallic core that the cladding material would be secured to when forming the finished fan blade. As the mandrel pressed into the blank, the result is shown in Figure 5.1B, in which the blank has been stretched and the walls of the mold have been adjusted to accommodate the length and curvature of the mandrel, thereby forming the cladding piece.

Teets presented the idea to Burnham, the General Manager of DRB, who recognized its potential but declined to pursue it since it would require altering the delivery schedule already in place for the cladding piece. Burnham therefore instructed Teets

---

[6] *J. Michael Teets v. Chromalloy Gas Turbine Corporation*, 83 F.3d 403 (Fed. Cir. 1996).

[7] "Hot Forming Process," United States Patent No. 5,694,683; Teets and Burnham, inventors; issued December 9, 1997.

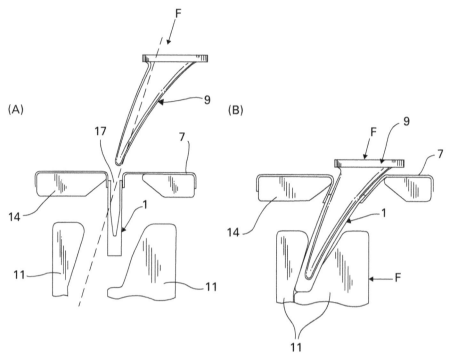

Figure 5.1   Selected figures from Patent No. 5,694,683. (A) Mandrel and mold cavity prior to molding operation. (B) Mandrel inside mold cavity after molding operation.

to focus on the welding process instead and to try and make it viable. Teets nevertheless worked on both processes, using DRB's materials and equipment and doing so on company time, and he and Burnham eventually obtained the patent through the company's patent counsel.

Burnham was obligated under an employment agreement to assign his inventions to DRB and did so in this case, but Teets had not signed such an agreement and did not assign his interest in the patent. Teets went to court to obtain a declaration of his ownership rights, but the court decided against him, noting that the specific goal of his project was to develop a one-piece cladding, which he had received direct instructions to invent as part of his job, and that the hot forming process was within the scope of what he was hired to invent.

When the patent application was filed, Teets executed the inventor's declaration that is a standard requirement of a patent application, but he refused to assign his share of the ownership of the patent application to Chromalloy. Since the application was filed prior to the enactment of the AIA, Chromalloy had to obtain a determination by a court of law of its right to Teets' share and hence full ownership of the patent. If the same facts were to occur now, Chromalloy would be able to inform the PTO directly of Teets' obligation to assign as one who was hired to invent, and if Chromalloy's claim was accepted by the PTO, the PTO would have granted the patent to Chromalloy without a court order.

While the Teets case involved an employee, it is likely that the same standard would be applied to an outside contractor whose agreed-upon duties clearly included solving a particular problem or developing technology in a particular area. Hence, an obligation to assign will likely be recognized outside of the employment context.

5.5 HAVING A "SUFFICIENT PROPRIETARY INTEREST" OTHER THAN BY ASSIGNMENT OR OBLIGATION TO ASSIGN

A company may also have a proprietary interest in an invention even if it has neither an assignment of the invention nor an obligation on the part of the inventor to assign the invention, and even if it is not an employer of the inventor. The AIA also allows such a company under limited circumstances to stand in, or substitute itself, for an unwilling inventor and apply itself for a patent. At the writing of this book, unfortunately, the PTO provides only limited guidance as to what these circumstances are. According to the AIA, the proprietary interest must be a "sufficient proprietary interest," and the PTO states[8] that the company's action in standing in for the inventor must also be "appropriate to preserve the rights of the parties." The terms "sufficient" and "appropriate" are not defined further, except to state[9] that the sufficiency of the proprietary interest and the appropriateness of the company's action to preserve the interest of the parties must be established by a statement of the facts forming the basis of the interest. The PTO advises that the statement be supported with a legal memorandum signed by a qualified attorney and indicating any statute, court decision, or other document that supports what the company claims to be a sufficient interest. One can speculate that a "sufficient proprietary interest" other than an assignment or obligation to assign might be an assignment of a part interest, a license or an agreement to grant a license, or other business relationship between the inventor and the company. In any case, this feature of the AIA may evolve with guidelines and rules that are more specific as parties invoke this feature in the coming years.

5.6 WHEN NO ASSIGNMENT, OBLIGATION TO ASSIGN, OR PROPRIETARY INTEREST: THE "SHOP RIGHT"

A shop right is the right of a company to practice an invention covered by a patent that it does not own, without paying royalties to the owner of the patent, if the invention in the patent was developed with the use of the company's resources, such as tools, machinery, materials, employees, or time. A lawsuit illustrating how a shop right can arise was one involving the Arkansas Power & Light Company (AP & L).

[8] "Changes to Implement the Inventor's Oath or Declaration Provisions of the Leahy-Smith America Invents Act," *Federal Register*, Vol. 77, No. 157, Rules and Regulations, pp. 48776–48826, United States Government Printing Office (August 14, 2012).

[9] USPTO Manual of Patent Examining Procedure §409.04(f).

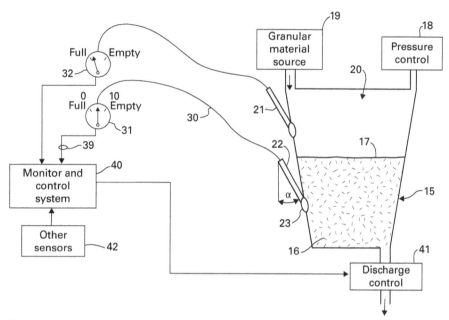

Figure 5.2   Figure from Patent No. 4,527,714.

AP & L generated electric power at a number of generating stations for sale to its consumers. One the stations was located near Redfield, Arkansas, and known as the White Bluff station. Power was produced at the White Bluff station from steam that was generated by coal-fired boilers that emit fly ash. Fly ash is a health hazard and requires disposal in an environmentally sensitive manner, and AP & L did so by using electrostatic precipitators and collecting the fly ash in the precipitator hoppers. For the installation, maintenance, and operation of the precipitators, AP & L engaged the services of an independent consultant. To prevent fly ash from escaping into the atmosphere, the level of ash collected in each hopper had to be monitored so that the electrostatic system for a given hopper could be turned off automatically once the ash reached a certain level. Level detectors were therefore incorporated in the system, but the detectors used were ones that relied on nuclear power, which itself raised environmental concerns and involved electricity that presented an explosion hazard within the hopper. Nonelectric sensors presented an alternative technology, one that relied on airflow and the pressure changes resulting from stoppage of the flow when the ash level reached the open end of the air line. These sensors became inoperative however when loose fly ash from the headspace above the ash level entered the air line and clogged the line. During the term of his consulting contract, the consultant conceived of a simpler level detector that depended on pressure changes but without air movement, using instead the pressure conditions that were normally encountered in the hopper operation. The concept is shown in Figure 5.2, which is a drawing from the consultant's Patent No. 4,527,714,[10] which became the subject of the lawsuit.

---

[10] "Pressure Responsive Hopper Level Detector System," United States Patent No. 4,527,714; Bowman, inventor; White River Technologies, Inc, assignee; issued July 9, 1985.

Referring to the figure, the pressure in the hopper (15) above the precipitated ash (17) was below atmospheric during normal hopper operation. The consultant's concept was to insert a pipe (21) in the hopper at the highest point that the fly ash level was permitted to achieve and to connect the pipe to a vacuum gauge (32). When the ash level became high enough to close the pipe opening, the vacuum gauge would show an increase in pressure above the normal subatmospheric level, which allowed an operator monitoring the gauge to take appropriate action, such as turning off the precipitator to prevent further fly ash from entering the hopper or drawing fly ash out from the bottom of the hopper by a discharge control (41) to lower the level.

During a power outage at the White Bluff facility, AP & L installed a detector based on the consultant's design on one of its precipitator hoppers to test out the concept, and when the test showed a positive result, AP & L installed the same detectors on additional hoppers, and ultimately on all of the hoppers at the station. All working drawings for the detectors were done at AP & L's expense, and AP & L also supplied all materials used in constructing and installing the detectors. Soon after all of the hoppers at the White Bluff stations were equipped with the new detectors, the consultant moved to another AP & L generating station, this one located near Newark, Arkansas, and known as the Independence Steam Electric Station (ISES), again to assist in the start-up, maintenance, and operation of precipitators at that location. A month later, while still under contract to AP & L, the consultant formed his own company, White River Technologies, Inc., to market his inventions and several months later filed a patent application on the level detector that had been installed earlier at White Bluff, the application eventually maturing into the patent cited earlier. Meanwhile, AP & L had White River Technologies install the level detectors on some of the hoppers at the ISES station but used another and less expensive contractor to install the rest. Again, all costs associated with the installation and testing of the detectors at ISES were covered by AP & L. Before installation and testing were complete, the consultant's contract expired. Two years after the expiration of the contract, an engineer at AP & L felt that the level detectors could also be useful on the hoppers fed by hydroveyors, of which there were several at the ISES facility. A hydroveyor, or water exhauster, creates a vacuum by directing high-pressure streams of water into a venturi throat. This causes air to be pulled through the venturi as well, and the resulting airflow pulls ash along with it in dry form. AP & L sent out a request for bids for outfitting the hydroveyor hoppers with the detectors, and White River Technologies responded but lost to a competitor. The detectors were installed, and all costs associated with the installation were covered by AP & L.

When the patent issued the same year, White River Technologies sued AP & L for infringement, and AP & L responded with the defense that it was entitled to a "shop right" to the invention and therefore not required to pay royalties to White River Technologies under the patent. White River Technologies argued that AP & L exceeded any shop right by putting the projects, including the level detectors that he had invented, out to bid and, in doing so, supplying details about the design and specifications to others outside AP & L. White River Technologies also argued that AP & L's duplication of the level detector on additional AP & L units without the consultant's involvement (after expiration of his contract) also placed AP & L's activities outside any shop right. The case ultimately went up on appeal, and the appeals court

Chapter 6

The Novelty Threshold: Can You Recognize It When You See It?

Chapter 2 sets forth what is or is not prior art. The next step is to compare the invention to the prior art in an evaluation of the patentability of the invention: Are there differences between the invention and the prior art that set the invention sufficiently apart to qualify the invention for a patent? The differences that qualify are those that make the invention "novel" and "nonobvious." Novelty requires that the invention differ in some way from the prior art, while nonobviousness requires that the difference be great enough to rise to the level of a patent grant. Admittedly, these explanations beg the question, but novelty and nonobviousness are distinct requirements, as evidenced by the fact that they are stated in separate sections of the patent statute, Sections 102 and 103, respectively, of Title 35, U.S. Code. The novelty section however is primarily a statement of what is or is not prior art, while the nonobviousness section does little more than state that an invention must be nonobvious over the prior art to qualify for a patent. As patent courts have been called upon to apply these two requirements to an untold variety of inventions, an intricate framework of definitions for each requirement has evolved. These frameworks are explored in this chapter and in Chapter 7, respectively.

As a general rule, however, an invention can only be evaluated for nonobviousness when it presents something novel, or at least different, to evaluate. Novelty is therefore a threshold requirement, one that an invention must meet before a patent examiner, a judge, or a jury will consider any arguments, explanations, test data, or evidence of any kind that a patent applicant, owner, or attorney might submit to convince the PTO or the courts of the nonobviousness of the invention. This discussion of patentability therefore begins with novelty. What makes an invention novel? And why do some inventors seem to be more prolific in producing novel ideas than others?

First to File: Patents for Today's Scientist and Engineer, First Edition. M. Henry Heines.
© 2014 the American Institute of Chemical Engineers, Inc. Published 2014 by John Wiley & Sons, Inc.

The answers to these questions begin with an understanding of what constitutes the process of inventing. Racking one's brain to think up novel ideas specifically for purposes of patenting or poring over the latest journal articles or the patents of others as a stimulus for generating ideas on one's own are neither necessary nor the best way to arrive at potentially patentable inventions. Novelty, and thus inventions that are candidates for patenting, can instead be found in any of the various adjustments, substitutions, variations, expansions, discoveries, and solutions that arise not only in addressing a problem but also in the performance of routine tasks, whether those tasks are assigned as part of one's job or simply encountered in everyday living. What the typical scientist or engineer may think of as a trivial change may indeed be a candidate invention for patenting. One of the keys to developing a patent portfolio is therefore the ability to recognize novelty, and patent law does so in a variety of ways, many of which might not be apparent to the casual researcher, technician, or problem solver, whether that person is an expert in the relevant field of technology or one still on a steep learning curve. The following sections describe some of the more prominent features of what the legal system recognizes as novelty in patent law and conversely what constitutes "**anticipation**," which is the absence of novelty.

6.1 ANTICIPATION AND THE "ALL ELEMENTS IN A SINGLE REFERENCE" RULE

It has often been said that all inventions are combinations of known elements, although many technologists will find it debatable as to whether a particular component or even the invention itself is an element or a combination. Nevertheless, many inventions are indeed mixtures or linked combinations of components or series or coordinated combinations of steps or functions, and one of the features of novelty in patent law is that these inventions are novel unless all of the components, steps, or functions are disclosed in a single piece of prior art. An example of an invention in which this question arose is the case of Broadcom Corporation's Patent No. 6,714,983.[1]

The invention in this patent is in the field of networks for receiving and transmitting radio-frequency (RF) communication data between mobile devices ("transceivers") and a data-processing terminal or between different mobile devices themselves or both. As examples of these mobile devices, the patent cites battery-powered, handheld devices such as those used in warehousing for inventory control and those used in rental car operations by van operators as they take orders from passengers on their way to the rental desk and by rental car lot attendants for logging cars and mileage as they are returned to the lot.

The filing date of the original application, which was refiled several times before Patent No. 6,714,983 was granted, was August 21, 1991, mentioned here to place the patent in its historical context in the development of mobile telephone technology.

---

[1] "Modular, Portable Data Processing Terminal for Use in a Communication Network," United States Patent No. 6,714,983, Koenck et al., inventors; Broadcom Corporation, owner; issued March 30, 2004.

The power sources available for mobile phones and similar devices in 1991 resulted in devices that were large and cumbersome compared to those being made decades later, and the invention in the patent reflected this. In addition, each business application had its own special requirements due to the different environments in which the various applications used the devices, the different distances over which they transmitted their signals, and the different frequency bands in which they were required to operate. As a result, each application had different RF communication characteristics, which in turn required different transceiver types and different antennae. In some applications, a single device had to support multiple sources of data, each with a different set of communication characteristics. Individual mobile devices could be made more flexible by using module cards for custom and application-specific functions and by designing the mobile device so that a module card could be inserted in a slot in the device, leaving the most basic functions to the device itself. The typical device had only one slot, however, and while devices existed that had two or more slots, these devices had to contain interfacing capabilities to accommodate the different communication characteristics received in the different slots. This interfacing required additional power, which made the multislot devices impractical for many portable applications. The invention in the Broadcom patent solved the problem by incorporating two or more communication transceivers in a single module that contained a communication processor and providing the module with the ability to roam between base stations at a controlled frequency to reduce the power needed by the communication processor and to communicate with different base stations in a structured manner.

When Broadcom asserted the patent against its competitor Qualcomm Incorporated, one of Qualcomm's defenses was that the patent was invalid since it was anticipated by (lacked novelty in view of) the Global System for Mobile Communications (GSM). The GSM was developed by the European Telecommunications Standards Institute (ETSI) and is a standard set of protocols for digital cellular networks used by mobile phones. The protocols are set forth in a set of specifications originally published by the ETSI the year before Broadcom's 1991 filing date. Although the Broadcom patent did not mention mobile phones, it claimed the invention in broad terms ("One or more circuits adapted for use in a mobile computing device comprising ... a terminal adapted to receive power for at least one of the circuits; ... communication circuitry comprising a reduced power mode and being adapted to use a first wireless communication and a second wireless communication ... to transmit data to access points, the ... circuitry reducing power by controlling the frequency of scanning for the access points; and processing circuitry ..."), and Qualcomm contended that the published specifications for the GSM collectively described a technology that met the language of the Broadcom claim. Qualcomm specifically cited 11 GSM specifications, even though Qualcomm's attorneys recognized that a claim is only anticipated if all elements of the claim are disclosed in a single reference document. Qualcomm's argument was that despite the fact that each of the 11 specifications was a separate document, the 11 specifications were like chapters of a book and functioned as a single, coherent reference. The case went through a trial and an appeal, and both courts disagreed with Qualcomm on this issue, the appellate decision[2] noting that the different specifications were authored by different subsets of authors at different times, each with its own subtitle and separate page numbering and each standing as a document in its own right. Since no single specification revealed all the elements of the Broadcom claim, anticipation was not present,[3] and the patent could not be declared invalid for lack of novelty.

When a single prior art document discloses less than all of the elements of a claim, the document can still however anticipate the claim if the missing element is "**incorporated by reference**" into the document from another document or if the missing element is deemed to be "inherent" in the document.

6.1.1   "Incorporation by Reference" of the Missing Element from Another Source

If a missing element is found to be "incorporated by reference" into a prior art document, the element is considered part of the document even though specific mention of the element is only found in another document. An example of an application where the invention was held to be anticipated by a single reference document that incorporated an element of the invention by reference to another document is Application No. 11/617,506 filed by Schlumberger Technology Corporation.[4] Drawings from the application are shown in Figure 6.1A and B.

The invention in the Schlumberger application was a device to be used in the manufacture of casings for oil and gas wells. The background of the invention was the known technique of first lining a wellbore (100 in Fig. 6.1A) with a casing (110) that is typically cemented in place, then perforating the casing at various points by detonating laterally directed explosive charges (125) on a perforating gun (120) that has been lowered into the wellbore. This created openings in the casing, with each opening extending laterally into the surrounding earth formation (105) to form elongated cavities (140) in the formation through which gas or oil from the formation could flow toward the casing openings and enter the wellbore. The purpose of the invention was to check the effectiveness of the perforating guns by determining the depths to which the cavities extended into the formation, and the invention achieved this by use of acoustic waves. To do this, an acoustic signal was generated at one depth inside the wellbore, and the resulting acoustic signal at a second depth was detected, the two depths arranged such that one (200 in Fig. 6.1B) was above and the other (210) below the depth of the cavity openings. The presence of a cavity caused the signal to become

---

[2] *Kyocera Wireless Corporation et al. v. International Trade Commission et al.*, 545 F.3d 1340 (Fed. Cir. 2008). Kyocera and the other appellants were Qualcomm's customers, and the ITC was involved because the patent owners claimed infringement of the patent by importation of the technology into the United States.

[3] Qualcomm also wanted to argue that the combined specifications made the claim obvious but, due to a procedural error, missed the opportunity to do so.

[4] "Tool for Measuring Perforation Tunnel Depth," United States Patent Application No. 11/617,506, filed December 28, 2006, published on May 10, 2007 as Publication No. US 2007/0104027 A1; Brooks, inventor; Schlumberger Technology Corporation, assignee.

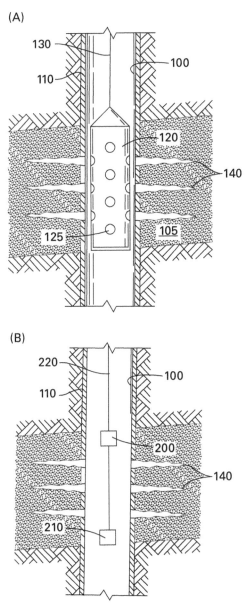

Figure 6.1   Selected figures from Patent Application Publication No. 2007/0104027. (A) Profile view of perforating gun. (B) Profile view of cavity depth measuring system.

attenuated by the Helmholtz effect between the generator and the detector, and the degree of attenuation was directly related to the depth of the cavity.

The claims in question in Schlumberger's patent application called for the use of acoustic transponders that contained piezoelectric transducers for the signal generator and the detector. These claims were rejected for lack of novelty over an earlier patent to Schlumberger's competitor Atlantic Richfield Company, Patent No. 5,218,573, that disclosed the same system but mentioned the acoustic signal transmitter and receiver only in general terms. The Atlantic Richfield patent did however state that the transmitter and receiver "may be similar to the type described in my U.S. Pat. No. 4,949,316" (another Atlantic Richfield patent), which indeed disclosed piezoelectric transducers although not for determining cavity depths. The context for the piezo-electric transducers in the '316 patent was in logging tools that are generally used for scanning a subsurface geological formation to provide information relating to the various strata and hydrocarbons that constitute the subsurface formation. Schlumberger responded to the rejection by arguing that the '573 patent did not itself disclose the use of piezoelectric transducers, but the PTO Appeals Board[5] maintained that the statement in the '573 patent was an incorporation of the '316 patent by ref-erence and that it was "a clear, detailed citation of what is being incorporated and where it can be found," so much so that "it is effectively part of [the'573 patent] as if it were explicitly contained therein." As a result, it was determined that the Atlantic Richfield patent that was the basis of the rejection (the '573 patent) disclosed by itself all elements of the claims in question in the Schlumberger application and thereby anticipated the claims. The Board therefore maintained the rejection, and Schlumberger ultimately dropped the application.

6.1.2   Inherent Disclosure of the Missing Element

The "**all elements**" **rule** can also be met by **inherent disclosure**, that is, an element that is not explicitly disclosed in a prior art document can be considered part of the document if the element is deemed to be inherently disclosed by the document. The question of what is or is not an inherent disclosure is illustrated by a case involving a patent whose novelty was challenged over two separate prior art references with opposite results.

   Patent No. 5,059,192[6] claims a method of hair removal using lasers. Since con-ventional methods of laser application to human skin tended to cause scarring or other damage to the skin, the invention in the patent lay in the "careful balance of laser parameters" to destroy the follicular papilla (the germ cells at the base of the hair follicle from which the hair grows) and not the surrounding tissue. The balanced parameters included using a laser with a wavelength that was selectively absorbed by the melanin in the papilla, placing the laser directly over the follicle opening and aligning the laser with the hair follicle, and applying the light in pulses whose inten-sity and duration caused damage to the papilla and not the surrounding tissue. The inventor, Dr. Nardo Zaias, got the idea while attending an American Academy of Dermatology trade show where Spectrum Medical Technologies, Inc., had a booth displaying its RD-1200 laser. While the display described the laser for use in removing tattoos, Dr. Zaias recognized that the principle involved in Spectrum's use

---

[5] *Ex parte* James E. Brooks, USPTO Board of Patent Appeals and Interferences (January 7, 2010).

[6] "Method of Hair Depilation," United States Patent No. 5,059,192, Zaias, inventor; issued October 22, 1991.

of the laser would work especially well when applied to the melanin in the papilla to permanently destroy the melanin and thereby prevent further hair growth. Dr. Zaias then proceeded to develop the hair removal idea and apply for the patent. The patent cited the Q-switched ruby laser as an appropriate laser for use in the invention, the laser producing light at the same wavelength as the RD-1200 laser that Spectrum was demonstrating at the symposium. When Spectrum eventually started selling the RD-1200 laser for use in hair removal, Dr. Zaias, together with his exclusive licensee Mehl/Biophile International Corporation, sued Spectrum for patent infringement.

Included among Spectrum's defenses to the lawsuit were challenges to the validity of the patent for lack of novelty over Spectrum's own manual for the RD-1200 laser that Spectrum had showed Dr. Zaias at the trade show and separately over an article published by another researcher in the field, Dr. Luigi Polla. Dr. Polla's article, entitled "Melanosomes are a primary target of Q-switched ruby laser irradiation in guinea pig skin," is directed to the selective destruction of pigmented cells by lasers and particularly to laser therapy for pigmented cutaneous lesions. The lawsuit ultimately reached the appeals court,[7] where the RD-1200 manual and the Polla article were considered separately. The court found that the RD-1200 manual described aiming the laser at skin pigmented with tattoo ink, without mentioning hair follicles, and the court observed that while tattoos could certainly extend over hair follicles, there was no necessary relationship between the location of a tattoo and the location of hair follicles. Noting that alignment of the laser with the hair follicle was an integral part of the Zaias invention as claimed, the court held that the manual did not disclose this alignment nor was the alignment inherent in any of the steps that the manual did disclose. Therefore, without either an actual or an inherent disclosure of each and every element of the claimed invention, the manual did not anticipate Dr. Zaias' claim. With the Polla article, however, the court reached a different result. Even though the article focused on epidermal lesions, it did refer specifically to hair follicles and to lesions that were located deep in the follicles and reported "follicular changes … associated with melanosome disruption" and more particularly "melanosomes contained within the follicular epithelium." The same laser that was used by Spectrum and highlighted in the patent was cited in the Polla article as well, and these disclosures, together with a description in the article of how the laser beam was aimed, led the court to conclude that the alignment feature of the patent claim was inherent in the description in the article. This meant that the article disclosed all parts of the claim, and the claim was therefore anticipated, despite the fact that the article's authors neither sought nor appeared to have appreciated the hair removal effect.

6.2   NOVELTY IN THE ARRANGEMENT OF PARTS

When all elements of an invention are disclosed in a single prior art document, the invention will still be novel if the arrangement of the elements in the invention differs from that in the document.

[7] *Mehl/Biophile International Corp., et al., v. Milgraum et al.*, 192 F.3d 1362 (Fed. Cir. 1999).

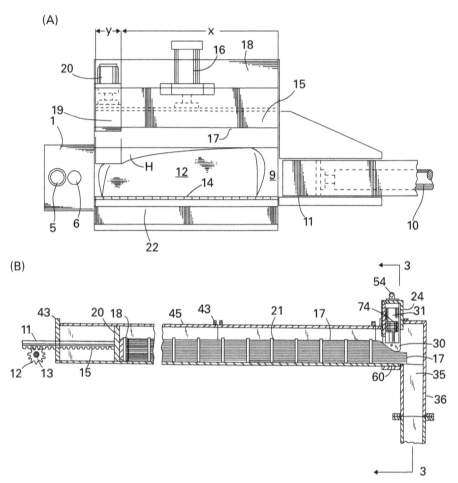

Figure 6.2   Selected figures from Patent No. 3,945,315 and prior art Patent No. 3,763,770. (A) Plan view from Patent No. 3,945,315. (B) Vertical section from Patent No. 3,763,770.

The elements listed in claim 1 of Patent No. 3,945,315[8] were the components of a hydraulic scrap-shearing machine designed to process both light-to-medium gauge metal objects and heavy gauge metal objects with a single hydraulic system. Figure 6.2A is a drawing from the patent.

The scrap-shearing machine has an open channel (9) through which scrap (12) is pushed by a ram cylinder (10) in the direction right to left, according to the view shown in the drawing. The channel (9) has a narrow mouth at its exit end (the left end), and scrap shears (not visible) are positioned just outside the narrow mouth. One side wall of the channel is formed by two "scrap-squashing" rams (15, 19) that are

[8] "Hydraulic Scrap Shearing Machine," United States Patent No. 3,945,315, Dahlem et al., inventors; Lindemann Maschinenfabrik GmbH, assignee; issued March 23, 1976.

independently movable but driven by a common hydraulic system, the working face of the downstream ram (19) having a smaller area than the working face of the upstream ram (15). When the scrap piece (12) occupies the length of the channel, both rams move together to "squash" the scrap against the opposing wall (14) of the channel, and when the scrap is light-to-medium gauge, the two rams squash it together. When the scrap is heavy gauge, however, the upstream ram (15) is not able to squash it and therefore stops moving. The downstream ram (19), on the other hand, can exert greater pressure because of its smaller working face and therefore continues to move, squashing the leading end of the scrap piece. Once squashed, the leading end can be advanced through the channel mouth and into the open shears, and the remainder of the scrap piece is weakened by the squashing of the leading end. Ultimately, the entire scrap piece is squashed to a diameter small enough to pass through the narrow mouth and into the shears. The elements of the machine as listed in the leading claim of the patent include the following: a channel with a movable side wall divided into two longitudinal sections of different lengths, a main hydraulic ram whose working face forms the longer section, and an auxiliary hydraulic ram whose working face forms the shorter section, the auxiliary ram being operable independently of the main ram, and finally, a hydraulic system for moving the movable side wall (i.e., the two rams) against an opposing side wall.

When the patent owner, Lindemann Maschinenfabrik GmbH, sued American Hoist and Derrick Company for patent infringement, American Hoist and Derrick responded that the claim was anticipated by an earlier patent, No. 3,763,770,[9] issued to Allied Chemical Corporation, a drawing from which is shown in Figure 6.2B. The Allied Chemical patent was directed to apparatus and method for cutting bundles of spent nuclear fuel rods into short lengths to make it easier to leach the rods in acid to extract the uranium from the rods. Admittedly, nuclear fuel rods and the bundles that they are mounted in differ considerably from heavy gauge scrap metal of the Lindemann patent, and the problems that the inventions in the two patents sought to address differ likewise. The Lindemann claim was not limited to use on metal scrap, however, but instead recited only the components and structure of the machine itself, as is typical with apparatus claims. The argument that American Hoist and Derrick made for anticipation was that, like the Lindemann patent claim, one of the structures shown in the Allied Chemical patent contained two independently actuated rams (referred to as "gags") of different-size working faces arranged side by side, an opposing surface (referred to as an "anvil") that the gags moved toward when a bundle was between the gags and the anvil, and a channel (referred to as a "magazine") through which the bundles were fed to the gags and anvil to be compressed laterally. In the view shown in the drawing, only one of the gags (30) is visible, since the gags were both directed at a single point along the axis of the magazine (the horizontal elongated channel shown in the drawing) and approached the magazine from orthogonal directions (i.e., at a right angle to each other). The anvil is likewise not visible but is located below the gag (30) and behind the bundle (17).

[9] "Method for Shearing Spent Nuclear Fuel Bundles," United States Patent No. 3,763,770, Ehrman et al., inventors; Allied Chemical Corporation, assignee; issued October 9, 1973.

The verdict reached at the trial court was in favor of American Hoist and Derrick, that is, holding that Lindemann's claim was anticipated by the structure shown in the Allied Chemical patent. The appeals court[10] however took a closer look at the Lindemann claim and the Allied Chemical structure and concluded that the arrangement of the parts in the Lindemann claim was different enough from the arrangement of the parts in Allied Chemical that anticipation was not present. Specifically, the court observed that the two gags in the Allied Chemical structure did not form a side wall of the magazine (as in the patent claim) since they were both positioned beyond the end of the magazine. Nor did they constitute a movable wall of the magazine since while the wall of the magazine had an adjustable width, this was achieved by a movable plate (parallel to the plane of the drawing and not shown) rather than the gags, which played no part in the adjustment. Nor was the anvil in the Allied Chemical patent positioned as a side wall of the magazine. The court concluded that "[t]he [trial] court's analysis treated the claims as mere catalogs of separate parts, in disregard of the part-to-part relationships set forth in the claims and that give the claims their meaning."

Inventions whose components are intangible rather than physical can also be novel by reason of a difference in the arrangement of their components. Examples are inventions involving the links between computers through the Internet. One such invention is that claimed in Patent No. 5,822,737,[11] which concerns credit card transactions conducted through the Internet and the need to provide consumers with security when purchasing goods online. Common means of online purchasing at the time involved four computers—those of the consumer, the merchant, the merchant's bank, and the bank issuing the credit card to the consumer. The invention in the patent introduced a fifth entity identified as a "payment processing" or "financial processing" entity to serve as a go-between among the other four. Use of the new entity involved downloading an encrypting software package to the consumer's computer so that the consumer's credit card number would be secure from hackers. A key claim in the patent recited a system containing five links between the various computers—one between the consumer's computer and a merchant's computer for communicating promotional material from the merchant to the consumer, a second between the consumer and the "payment processing" computer for communicating the consumer's credit card information and the purchase amount to the payment processing computer, a third between the payment processing computer and the computer of the consumer's bank for communicating the credit card information and amount to the consumer's bank and seeking authorization, a fourth between the payment processing computer and the consumer's computer for communicating transactional indicia, and a fifth between the payment processing computer and the merchant's computer for communicating the same transactional indicia. The indicia, according to the invention, are generated by the payment processing computer in a form specific to each transaction, and each link is terminated before another link is made.

---

[10] *Lindemann Maschinenfabrik GmbH v. American Hoist and Derrick Company et al.*, 730 F.2d 1452 (Fed. Cir. 1984).

[11] "Financial Transaction System," United States Patent No. 5,822,737, Ogram, inventor; issued October 13, 1998.

When the patent was asserted in an infringement suit, this key claim was challenged as being invalid due to anticipation by a document published earlier by the Internet Engineering Task Force and IBM, entitled "Internet Keyed Payments Protocol" (which the parties referred to as "the iKP reference"). The iKP reference had been generated by representatives of the Internet industry who, like the inventor in the patent, sought to provide ways for electronic payments to be made over the Internet in a secure manner. Unlike the patent, however, the iKP reference sought to use the existing financial infrastructure for authorization and clearance of the payments, without adding additional entities. The reference presented two standard models, both involving a sequence of communications between the consumer, the merchant, the merchant's bank, and the consumer's bank, regarding the item to be purchased, the purchase price, the consumer's credit card information, and authorizations to proceed, the two models differing in terms of which communication is exchanged between which entities. Despite the reference's limitation to the use of the existing financial network, various steps in the two models were compared with each of the five links in the patent claim, although to achieve the claimed set of links, one would have to select individual steps from the two models and recombine them. The trial court found the claim anticipated nevertheless, but the appeals court disagreed,[12] stating that the reference might have anticipated a claim directed to either of the two model protocols that the reference disclosed, but it could not anticipate the system claimed in the patent. The claim was therefore declared novel.

6.3 ANOTHER ARGUMENT AGAINST ANTICIPATION: THE "NONENABLING REFERENCE"

A prior art document can reveal an invention or even the elements of an invention in the correct arrangement and still not anticipate the invention if the description in the document is not "enabling." To be enabling in this context means to place the invention in the possession of the public by including enough description that the reader could reproduce the invention successfully from the description. A reference that is lacking in enablement will not support a rejection of a claim for lack of novelty.

Most of the cases in which the nonenablement argument has succeeded have involved inventions in the field of organic chemistry, particularly those claiming new chemical compounds. Courts have held that a reference disclosing a compound is not an enabling reference and therefore fails to negate the novelty of a later claim to the compound, if the reference merely shows the compound's structural formula without also describing how the compound can be made. An often-cited case[13] is one in which the inventor's claim to new compounds was rejected over a reference that described the synthesis of a number of compounds, all except two of which were not covered by the inventor's claim. Of the two that were covered, the reference indicated that the authors' attempts at synthesis were unsuccessful. The examiner recognized this and

[12] *Net MoneyIN, Inc. v. VeriSign, Inc.*, 545 F.3d 1359 (Fed. Cir. 2008).

[13] Application of Wiggins et al., 488 F.2d 538 (C.C.P.A. 1973).

addressed it by citing a second reference, dated 2 years after the first (although still well before the inventor's filing date), showing a different but successful synthesis process. The second reference did not disclose any of the compounds that the inventor was claiming, but the process that it disclosed was similar to the one the inventors had used successfully in preparing the compounds they were now claiming. The examiner reasoned that since the successful synthesis method was in the second reference and hence in the prior art, the skilled synthesis chemist would know how to make the two compounds shown in the first reference despite the lack of success in that reference. When the inventor challenged the examiner in court, the court held that the listing of the compounds by name in the first reference "constituted nothing more than speculation about their potential or theoretical existence. The mere naming of a compound in a reference, without more, cannot constitute a description of the compound, particularly when, as in this case, the evidence of record suggests that a method suitable for its preparation was not developed until a date later than that of the reference."

Another case[14] involving organic chemistry reached a similar result but only after the inventor's attorney submitted an affidavit by an expert. The reference in this case disclosed compounds within the scope of the invention but did not report an attempt to make them, even though it reported a process for making other compounds that were not within the scope of the invention. The affidavit stated that the process that was included in the reference would not work for the compounds that the inventor was claiming and that there were no processes that were publicly available at that time that would work, the inventor being the first to achieve success. The court agreed, overturning the examiner, and stated that for a reference to defeat the novelty of an invention, "its [the reference's] disclosure must be such that a skilled artisan could take its teachings in combination with his own knowledge of the particular art and be in possession of the invention."

Defeating a reference for lack of enablement can also occur in inventions other than those claiming new chemical compounds, although most attempts to do so do not succeed. Those that do tend to be cases where the argument is supported by expert testimony similar to the Hoeksema case. While the expert in that case was someone other than the inventor, the inventor can also be established as an expert and can submit a declaration to the same effect.

One example is Patent No. 7,521,082[15] in which the invention is a process for forming a high-temperature superconducting (HTS) medium as a thin film on a metallic substrate. The compositions of the HTS film and substrate were both known, but the characteristic features of the process included reacting raw materials in the vapor phase above the substrate and depositing the product on the substrate in a process known as chemical vapor deposition (CVD) to form a superconducting thin film having a critical current and a current density above certain stated levels at a

[14] Application of Hoeksema et al., 399 F.2d 269 (C.C.P.A. 1968).

[15] "Coated High Temperature Superconducting Tapes, Articles, and Processes for Forming Same," United States Patent No. 7,521,082, Selvamanickam, inventor; Superpower, Inc., assignee; issued April 21, 2009.

temperature of 77 K and a condition known as self-field. During examination of the application for this patent at the USPTO, some of the claims were rejected as anticipated by a patent to Arendt et al.,[16] which described depositing a thin film of superconductor material on a variety of substrates including metal substrates. The Arendt patent also listed a variety of superconductor materials and a variety of deposition methods, including CVD, but the working examples in the patent used only pulsed laser deposition (PLD). The patent attorney responded to the rejection by submitting a declaration (sworn statement) by the inventor himself, listing the inventor's credentials, expertise, and 12 years' experience in research relating to high-temperature superconductors. The declaration stated that Arendt included CVD only as part of a "laundry list of deposition techniques" and that Arendt "is no more relevant than the prior art of which I was aware ... and is limited to the achievement of high quality superconducting thin films by physical vapor deposition, particularly PLD." The declaration further stated that while "there is a general interest in the art of creating high-quality superconducting thin films by chemical vapor deposition, the art has failed to even remotely enable such formation" and that he was able to achieve a successful CVD process only through "exhaustive research and development efforts over an extended period of time," referring to his work on the apparatus and operating conditions that he ultimately used. While the declaration failed to convince the examiner, the Appeals Board reversed[17] the examiner and found for the inventor, holding that "Arendt did not put the claimed method in possession of those of ordinary skill in the art, i.e., enable those of ordinary skill in the art to form, by CVD, a thin film having the claimed critical current and current density properties."

6.4 CAUTION: A REFERENCE CAN ANTICIPATE AN INVENTION EVEN IF IT "TEACHES AWAY" FROM THE INVENTION

Many inventors have attempted to argue that their inventions are not anticipated by references that disclose their inventions but do so in a manner that "teaches away" from them, that is, references that in some way disparage the ideas that the inventors later use as the basis for their inventions. Unfortunately, these arguments do not succeed.

One example occurred in Patent Application No. 10/193,008[18] of inventor Meggiolan on a device for closing a bicycle chain (the drive chain joining the pedals to the rear wheel) by linking the two ends of the chain together to form a loop. A drawing from the application is presented in Figure 6.3.

---

[16] "High Temperature Superconducting Thick Films," United States Patent No. 5,872,080, Arendt et al., inventors; The Regents of the University of California, assignee; issued February 16, 1999. The distinction between "thin" and "thick" films did not arise in either the declaration or the Board's opinion.

[17] *Ex parte* Venkat Selvamanickam, USPTO Board of Patent Appeals and Interferences, June 25, 2008.

[18] "Device for Closing a Chain, Particularly a Bicycle Chain," Application No. 10/193,008, filed July 11, 2002, Meggiolan, inventor. The application, which was ultimately abandoned, was published as Publication No. US 2003/0022748, publication date January 30, 2003.

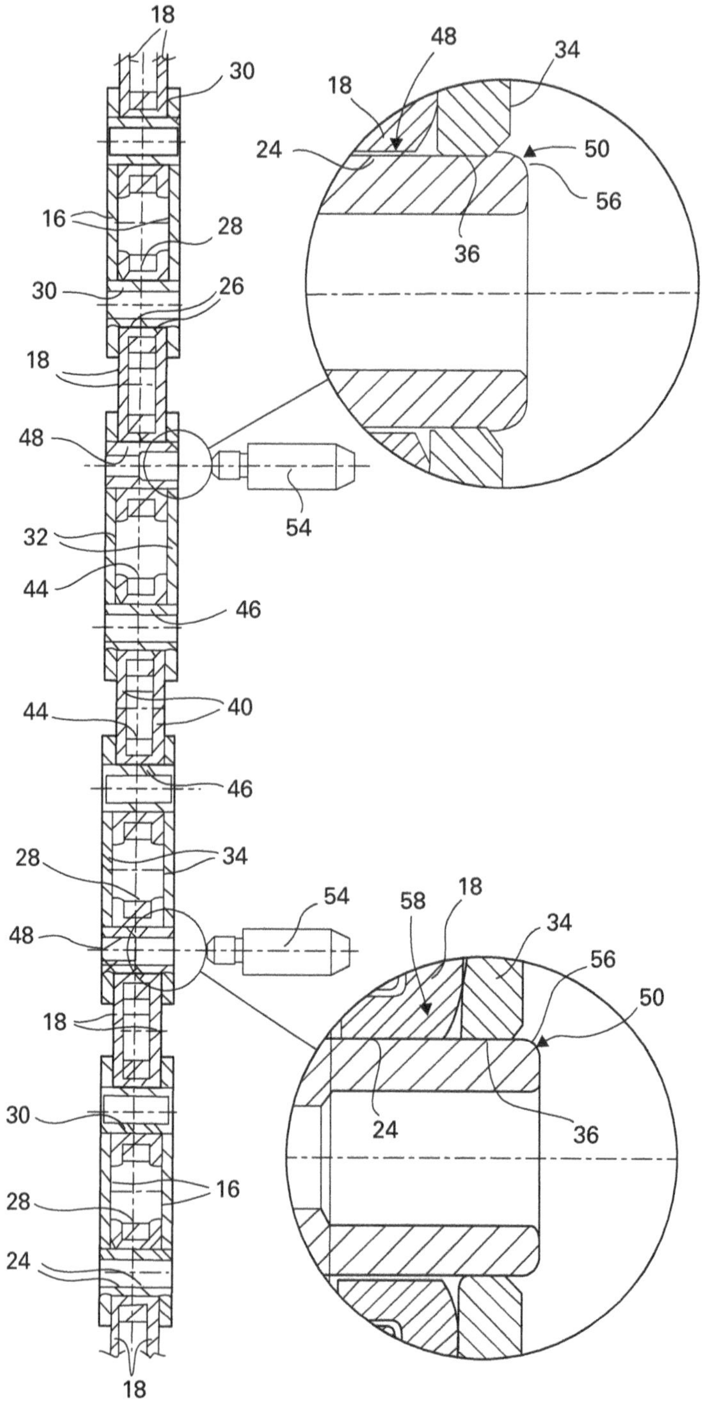

Figure 6.3   Selected figure from Patent Application Publication No. 2003/0022748.

Meggiolan's invention arose because chain lengths often need adjustment to fit bicycles of different sizes and also because bicycle gears or sprockets occasionally require replacement, and this likewise entails opening and closing the chain. As bicycle riders well know, a bicycle chain consists of parallel pairs of oblong plates of a relatively wide spacing (16 in the figure) alternating with parallel pairs of plates that are closer together (18), each wider-spaced pair being joined at each of its two ends to the ends of a narrow-spaced pair in a pivot joint that gives the chain its flexibility. The pivot joint includes a rivet (30) that passes through aligned holes in both pairs of plates, and the problem that the invention was addressing was that, because of the way that the rivets were secured to the plates, the removal of a rivet to open the chain tended to cause damage to one or both of the outer plates. The chain in the Meggiolan invention featured a rivet that terminated in a "clip engagement portion" (50) that "elastically widened" the hole (36) in one of the outer plates. When the rivet was inserted first through the other three plates and then forced through the "elastically" widenable hole in the fourth plate, the clip engagement portion of the rivet engaged the rim of that hole, causing the rivet to securely join all four plates in a way that allowed them to pivot but not bend. Although not stated in the Meggiolan application, the clip engagement portion apparently also worked in reverse, leaving the "elastically" widenable hole undamaged upon both insertion and removal of the rivet.

Although the "clip engagement" construction was shown in the drawings and described in the text of the application, the claims presented the invention more broadly, as a combination of pairs of inner plates and outer plates and a pair of rivets for connecting end holes in the various plates. As a result, the application was rejected as anticipated by the description in a patent to Ta Ya Chain Co, Ltd.,[19] that issued several years before. Drawings from the Ta Ya patent are shown in Figure 6.4A and B.

The Ta Ya patent addressed the same problem addressed by the Meggiolan application but proposed a different solution: the holes (8a, 8b) in Ta Ya's outer plate were compound holes formed by two eccentric but overlapping circles, one with a smaller diameter than the other, and the rivets (28a, 28b) that joined the outer plates to the inner plates each had a wide head (32a, 32b) at its leading end and a relatively narrow neck (30a, 30b) connecting the head to the shaft of the rivet. The head could only pass through the large diameter section of the hole, while the neck could pass through either section, including the small diameter section. To insert the rivet, the parts of the chain to be joined were held at a slight angle to each other to allow the head to pass through the wider portion of the hole, and when the chain was then straightened, the head shifted into position over the narrow portion, thereby securing the rivet in place. To achieve the angle needed for connecting and disconnecting, therefore, the chain had to be bent slightly out of the plane of the plates, while the rivet in the Meggiolan patent application could be installed without bending the chain. Meggiolan's attorney responded to the rejection by arguing that the Ta Ya patent "teaches away from Applicant's configuration. It is the rigidity of the adjustable links

(A)

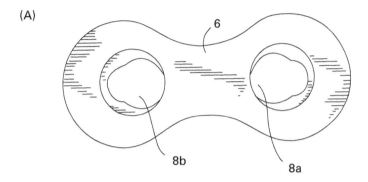

(B)

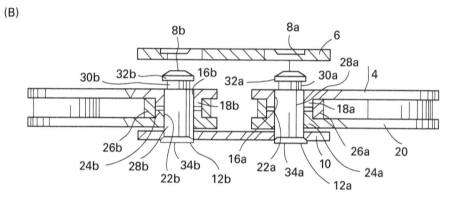

Figure 6.4   Selected figures from Patent No. 5,178,585. (A) Plan view of outer chain plate. (B) Cross section of links prior to joining.

that permits their easy removal and replacement. ... [I]n order to remove or replace the outer plate [in the chain in the Ta Ya patent], ... the chain is bent so that the connecting pins are no longer parallel ... Were these characteristics to be imparted in a 'chain closing device' of multiple links, rigidity would be lost due to bending of the chain links with respect to each other, making removal and replacement of the chain plates more difficult." When Meggiolan's attorney took the application up on appeal, the PTAB[20] rejected the argument, stating that "[t]he question whether a reference 'teaches away' from the invention is inapplicable to an anticipation analysis."

In another application, the inventor used the "teaches away" argument over a reference that actually described his invention but did so in a negative way, and the argument faced the same obstacle. The technology addressed by U.S. Patent Application No. 10/788,979[21] (Applied Materials, Inc.) was integrated circuit

[20] *Ex parte* Mario Meggiolan, United States Patent Application No. 10/193,008, Decision on Appeal, February 1, 2013.

[21] "Backside Rapid Thermal Processing of Patterned Wafers," United States Patent Application No. 10/788,979, Aderhold et al., inventors; Applied Materials, Inc., assignee; published as Publication No. US 2005/0191044, publication date September 1, 2005.

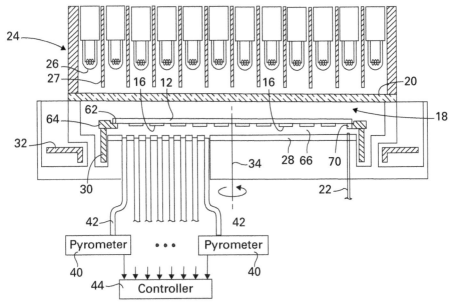

Figure 6.5 Selected figure from Patent Application Publication No. 2005/0191044.

fabrication and its use of rapid thermal processing (RTP). A drawing from the published version of the application appears in Figure 6.5.

RTP was performed by radiation heating, and the need for highly controlled and uniform heating over two-dimensional areas was increasing as the circuit features on the wafers became smaller and the wafers themselves became larger. Various features were introduced to address these needs, including zoned heating and a series of pyrometers distributed across the zoned area to provide dynamic control of the heating. In the typical arrangement, the radiation heat source was located above the wafer, directly facing the circuit features, and the pyrometers were located below, facing the back side of the wafer. The invention lay in reorienting the wafer so that the radiant heat source was directed to the back side of the wafer and the pyrometers were on the front side. In the drawing, the wafer (12) is shown upside down with its features (16) exposed on the bottom. High-intensity tungsten-halogen lamps (26) serve as the radiant heat source and are positioned above the inverted wafer, and the pyrometers (40) are positioned below, receiving signals through light pipes (42).

The reference cited against the application was a patent to Texas Instruments (TI),[22] in which the same problem was addressed. The solution in the TI patent was to first generate calibration curves using a reference wafer having a known reflectivity, by determining both the amount of radiation emanating from the heat lamps and the amount reflected from the wafer to obtain real-time values for both emissivity and temperature. The figures in the patent showed the wafer face down with the heat

---

[22] "Method for Temperature Measurement in Rapid Thermal Process Systems," United States Patent No. 5,601,366; Paranjpe, inventor; Texas Instruments Incorporated, assignee; issued February 11, 1997.

source above, as in the claims of the Applied Chemicals application, and the text of the patent mentioned the use of standard pyrometers without showing them in the figures. The text included the following statement:

> *While the pyrometer can be used to image the radiation from either surface of the wafer, higher accuracy can be obtained for the unpatterned wafer backside.*

The argument made by Applied Materials was that the statement, and hence the patent, "teaches the unclaimed combination of backside pyrometric monitoring and topside irradiation of an inverted wafer ... [and] further teaches against frontside pyrometric monitoring of a wafer of whatever orientation." The PTAB heard this argument on appeal[23] and responded by pointing to the phrase "from either surface of the wafer" in the TI patent. The PTAB concluded that despite the negative light in which the TI patent referred to the combination of backside irradiation and frontside pyrometric monitoring, the patent disclosed the combination nevertheless and observed that "the subject matter of a claim 'does not become patentable simply because it has been described as somewhat inferior to some other [alternative] for the same use'."

6.5   NOVELTY VERSUS ANTICIPATION AMONG GENUS, SUBGENUS, AND SPECIES

The most intricate and least apparent part of the novelty framework may be in the rules governing genera, subgenera, and species. Inventions that lie in the discovery of a genus, for example, may confront prior art in the form of a species within the genus, while inventions that lie in the discovery of a species may confront prior art in the form of a genus that includes the species. The discovery of a subgenus can raise similar issues. Which of these discoveries constitutes a novel invention?

6.5.1   Species Anticipating a Genus

Genus-, subgenus-, and species-type inventions typically arise in chemistry, and a claim to a genus is often made when members of the genus, or at least representative members, are discovered to have a special property or utility that was previously unrecognized. If any member of the genus is already in the prior art, the genus lacks novelty even though it may never have been recognized as a genus or expressed as such. An example is the case of *In re* Gosteli et al., 872 F.2d 1008 (Fed. Cir. 1989), where the invention lay in chemical compounds as intermediates to certain antibiotics. The application contained claims to a genus of the compounds specified by a generic formula, together with narrower claims to a group of 21 specific compounds within the genus. A reference disclosed two compounds within the genus, one of which was not included among the application's 21 examples and the other being one of the 21. The reference also listed other compounds that were outside the genus. The

---

[23] *Ex parte* Aderhold et al., United States Patent Application No. 10/788,979, Decision on Appeal, May 1, 2013.

reference did not disclose the genus but was nevertheless held to anticipate the claim to the genus and a claim to the specific compound.

In this case, the utility stated in the reference for the compounds was the same as that stated in the application for the rejected claims, that is, both presented the compounds as intermediates for the same type of antibiotics. Even if the utilities were unrelated, however, anticipation would be present, since the claims were to the compounds themselves, expressed either generically or as individual compounds.

6.5.2 Specific Value Anticipating a Range

Analogous to a species anticipating a genus, a single value can anticipate a range that encompasses the value. In *Titanium Metals Corp. v. Banner*, 778 F.2d 775 (Fed. Cir. 1985), the claims at issue were directed to a titanium alloy specifying a nickel content within the range of 0.6–0.9% and a molybdenum content within the range of 0.2–0.4%. The reference cited against the claims included a graph showing a specific data point of a titanium alloy with 0.25% nickel and 0.75% molybdenum, which was held to anticipate the claims that listed the ranges. Other claims in the same application recited specific values rather than ranges, and one claim was limited to 0.8% nickel and 0.3% molybdenum. Since no alloy with these particular nickel and molybdenum levels was actually shown in the reference, this claim was deemed novel.

6.5.3 "Shotgun" Disclosures in the Prior Art

A "shotgun" disclosure is one that lists the species that one is attempting to claim among a large number of other species, and inventors are tempted to argue that the disclosures of this type should not be considered anticipation since one would have to know to look specifically for the species in order to find it in the list. This argument typically fails to succeed. An example is the case of an application on a moldable polycarbonate resin invented by Sivaramakrishnan.[24] The invention lay in enhancing the hydrolytic stability of the resin by including either cadmium 2-ethylhexanoate or cadmium laurate in the resin. The reference disclosed polycarbonate resins whose flame resistance was enhanced by the inclusion of metallic salts, and the reference listed two full columns of examples of metallic salts, both specifically and by generic formulae, with cadmium laurate (but not cadmium 2-ethylhexanoate) explicitly included in the list. The inventor's attorney argued against the reference, stating that it gave no indication of anything special about cadmium laurate, and submitted comparative test data showing the superiority of cadmium laurate over other metal salts listed in reference to support the argument. The court still held that reference anticipated the claim.

The patent was granted however[25] when the claim was amended to restrict the polycarbonate in the resin to one that was halogen-free, since the polycarbonates in the reference were all halogenated.

[24] *In re* Sivaramakrishnan, 673 F.2d 1383 (CCPA 1982).

[25] "Polycarbonate Having Improved Hydrolytic Stability," United States Patent No. 4,374,226; Sivaramakrishnan, inventor; Mobay Chemical Corporation, assignee; issued February 15, 1983.

6.5.4   Species or Subgenus Novel over a Larger, Encompassing Genus

The obverse of the question of a claimed genus versus a prior art species is that of the claimed species versus a prior art genus, and many inventors will be surprised to know that the result is the opposite, provided that the prior art does not explicitly disclose the species in addition to the genus. An example is the case of a patent directed to a process for 2-nitrobenzaldehyde,[26] where the claim recites treating an alkali metal salt of the corresponding 2-nitropyruvic acid with an "alkali metal hypochlorite." The application on which the patent was granted had been rejected over a reference disclosing the same process except for the use of an "alkaline chlorine or bromine solution" rather than an alkali metal hypochlorite. Both "alkali metal hypochlorite" and "alkaline chlorine or bromine solution" are generic expressions, although "alkali chlorine or bromine solution" is a much larger genus. The reference also listed sodium hypobromite and calcium hypochlorite as specific examples, but no alkali metal hypochlorites (calcium is not an alkali metal), and for this reason, the claim was held to be novel.[27]

6.5.5   Narrow Range Novel over a Broad Range Encompassing the Narrow Range

Cases that are parallel to those involving a novel chemical species or a novel subgenus of chemical compounds are those involving a parameter such as pressure, temperature, dosage, humidity, length of time, etc., expressed as a specific value or a narrow range and rejected over prior art disclosing the same parameter but in a broader range. A pressure range was the focus of a challenge to a patent on fluorescent light bulbs.[28] As claimed in the patent, the bulbs contained, among other components, a buffer gas at a pressure "less than 0.5 torr" in combination with a power source that produces a high discharge current in the lamp envelope. The challenge was by an infringer who asserted that the claim was anticipated by a reference that disclosed a fluorescent bulb with the same components but listed the buffer gas pressure as "1 torr or less." Despite the closeness of the ranges, the appellate court declared the claim novel in view of its narrower range.[29]

---

[26] "Process for the Preparation of 2-Nitrobenzaldehyde," United States Patent No. 4,203,928; Meyer, inventor; Bayer Aktiengesellschaft, assignee; issued May 20, 1980.

[27] In the Matter of the Application of Horst Meyer, 599 F.2d 1026 (Fed. Cir. 1979).

[28] "High Intensity Electrodeless Low Pressure Light Source Driven by a Transformer Core Arrangement," United States Patent No. 5,834,905; Godyak et al., inventors; Osram Sylvania Inc., assignee; issued May 10, 1998.

[29] *Osram Sylvania, Inc. v. American Induction Technologies, Inc.*, 701 F.3d 698 (Fed. Cir. 2012).

6.6   ARE WE DONE?

The list of comparisons presented in this chapter indicating what is considered novel and what is considered anticipated is not comprehensive—additional arguments, for and against novelty, have been made and passed on by the PTO and the courts. For example, a common mistake is to argue in favor of a machine, apparatus, device, or composition of matter by stressing that its intended use is different from that of the prior art. An intended use that is novel will not make the machine, apparatus, device, or composition of matter novel unless the novel use is somehow incorporated into the claim language to limit the structure or composition of the invention as claimed. This is but one example of overcoming a rejection for lack of novelty by narrowing the scope of the rejected claim, and this can often be done without a significant loss of coverage to the patentee. Claims can also be recast to avoid a novelty rejection in ways other than simply narrowing the scope. Many claims to articles of manufacture or compositions, for example, can be recast as claims to a process for manufacture or a method of use, that is, a method for achieving a result or effect by utilizing the article or composition.

Novelty is not a question of degree: an invention is either novel or anticipated, and if it is anticipated, the door to patentability of that invention is closed. As noted, however, novelty is only the first part of the patentability inquiry. As the next chapter will demonstrate, the difference that causes an invention to be novel can be great enough on its own that it also meets nonobviousness, and often, an explanation to that effect will suffice. In other cases, more is needed.

Chapter 7

Confronting the Prior Art: What Makes an Invention Nonobvious?

The word "nonobviousness" is a clumsy combination of word segments and lacks a precise meaning; attempts to define it have produced a body of law that can fill volumes. To add to the confusion, it's a negative word that expresses a positive conclusion (a nonobvious invention is a patentable invention), rather than the features of an invention that lead to the conclusion. Those consulting the patent statute for its expression of the nonobviousness requirement will find that the statute simply says that to be patentable an invention must be "[nonobvious] to a **person having ordinary skill in the art** [often shortened to the acronym 'PHOSITA'] to which the claimed invention pertains."[1] The statute thus fails to state the skill level of the PHOSITA, leaving this task to the PTO and the courts, which have taken widely varying approaches in their attempts to identify it. In biotechnology inventions, for example, some cases state that the PHOSITA must have not only a doctoral degree but postdoctoral research experience[2]; others state that "a bachelor's degree in pharmaceutical science or analytical chemistry, and some experience in drugs and drug preparation"[3] are sufficient. At least one case states that "the hypothetical person is not definable by way of credentials,"[4] and many highly skilled workers know that in almost any field of expertise, there are individuals who despite lacking a college degree and in many cases even a high

---

[1] 35 U.S.C. §103, the quoted wording virtually unchanged by the AIA.

[2] *Bd. of Trs. of Leland Stanford Junior Univ. v. Roche Molecular Sys.*, 563 F. Supp. 2d 1016, 1034 (N.D. Cal. 2008).

[3] *Pfizer, Inc. v. Apotex, Inc.*, 480 F.3d 1348 (Fed. Cir. 2007).

[4] *Ex parte* Hiyamizu, 10 USPQ2d 1393, 1394 (USPTO Board of Patent Appeals & Interferences 1988).

First to File: Patents for Today's Scientist and Engineer, First Edition. M. Henry Heines.
© 2014 the American Institute of Chemical Engineers, Inc. Published 2014 by John Wiley & Sons, Inc.

school diploma have through experience acquired skill levels that are well above what might be considered the "ordinary skill in the art." The PHOSITA is nevertheless at the core of nonobviousness evaluations, and nonobviousness itself is possibly the requirement that has developed the most intricate set of rules to determine what is needed for it to be met.

What is clear, as noted in Chapter 6, is that whether an invention does or does not meet the nonobviousness requirement is an inquiry that only applies to inventions that meet the novelty requirement and that nonobviousness determinations are evaluations of the quality of the differences that make inventions novel. In their attempts to identify this quality, the courts for a time expressed the view that an invention had to entail a "flash of genius,"[5] as opposed perhaps to being the result of mere tinkering. The "flash of genius" standard proved to be unworkable, however, and was overruled by Congress in the 1952 Patent Act, which explicitly stated that "Patentability shall not be negatived by the manner in which the invention was made."[6] Nonobviousness is thus determined by an evaluation of the invention itself, including its properties, operability, and utility, not by how much brainstorming, testing, or searching the inventor had to go through to arrive at the invention.

Still, however, one is left without a definition, although certain tests have been articulated, one of which was the **"teaching, suggestion, or motivation" (TSM) test**. According to the TSM test, an invention is only obvious if the prior art, the nature of the problem, or the knowledge of a PHOSITA reveals some motivation or suggestion to combine the teachings of the prior art in a way that results in the invention. The Supreme Court confronted this test in 2007 in the pivotal and widely cited case of *KSR International Co. v. Teleflex Inc. et al.* and overruled it, stating that while it is one means of evaluating an invention for nonobviousness, it is not the only means. The Court stated that "[a] person or ordinary skill is also a person of ordinary creativity, not an automaton," and that "[r]igid preventative rules [such as the TSM test] that deny factfinders recourse to common sense … are neither necessary under our case law nor consistent with it."[7] Neither this statement nor others in the Court's decision offer much help. Rather than offering positive guidelines to inventors, the KSR decision is a loosening of otherwise rigid rules and the granting of greater freedom to the USPTO and the lower courts to declare an invention obvious and hence unpatentable.

Nevertheless, certain approaches to evaluating inventions for nonobviousness, some of which are specialized to certain technologies, have been developed by the USPTO and the courts and have continued to be applied within the guidelines of the KSR decision. The following sections address the more common approaches.

---

[5] *Cuno Engineering Co. v. Automatic Devices Corp.*, 314 U.S. 84 (1941).

[6] 35 U.S.C. §103.

[7] *KSR International Co. v. Teleflex Inc. et al.*, 550 U.S. 398 (2008).

7.1 "BUT EVERY INVENTION IS A COMBINATION OF OLD ELEMENTS!"

While this maxim was mentioned in the discussion of novelty in Chapter 6, it tends to be invoked with greater frequency in arguments over **obviousness**. Surely, every invention can be broken down into parts, separable or not, and while the question of whether a particular part is new or old may depend on how the "part" is defined, any part or subpart can be further broken down until it is reduced to a combination of elements that are somewhere in the prior art. Nevertheless, there are a number of ways in which combinations of known elements can be shown to be nonobvious.

7.1.1 Synergism and Changes in Function

The question of whether a combination is obvious or nonobvious often comes down to asking whether the individual elements when combined function any differently than they did individually.

A case in which a combination was deemed obvious because all of its elements performed the same function that they performed in the prior art is the case of Sundance, Inc.'s Patent No. 5,026,109,[8] a drawing from which is shown in Figure 7.1.

The invention in the Sundance patent was directed to retractable cover systems used as covers for truck and trailer bodies or for swimming pools or as awnings for

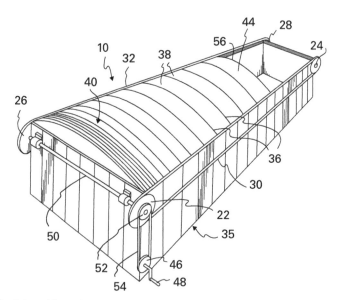

Figure 7.1   Selected figure from Patent No. 5,026,109.

---

[8] "Segmented Cover System," United States Patent No. 5,026,109, Merlot, inventor; Sundance, Inc., assignee; issued June 25, 1991.

porches and patios. When a tear or hole occurred in one of these covers or awnings, replacement of the entire cover system seemed unnecessarily expensive. The invention therefore resided in a cover or awning constructed in sections, each of which could be removed and replaced independently of the others. The elements of the invention included individually removable cover sections (44); a series of parallel supporting bows (36), that is, rods running along the edges of the sections between each adjacent pair of sections, with the edges of the sections attached to the rods in a detachable manner; parallel cables (30, 32) to hold the ends of the bows while allowing them to slide as the sections were extended or retracted; and a drive assembly (48) to extend or retract the entire set of sections. The drive assembly operated by moving the bow at the far end of the set, according to the view shown in the drawing, while the bow at the near end was fixed.

When Sundance sued its competitor DeMonte Fabricating Ltd. for infringement, DeMonte responded that the patent was invalid for obviousness, citing two older patents as prior art. The first of the two older patents was directed to "a tarpaulin cover system for use in trucks" that included all of the elements of the Sundance cover system except the segmented cover. The second patent was directed to "a flexible canopy structure," also for use as "an expandable cover for trucks" among others, and it showed a segmented cover whose segments could be separated and exchanged, but it lacked some of the elements shown in the first patent. Separation of the segments was more complicated in the "flexible canopy structure" patent than in the Sundance patent, since each pair of adjacent segments in the canopy structure was joined by a face plate that had to be removed before the segments could be separated, and the removal of the face plate left the two segments hanging down rather than stretched over the canopy. Sundance argued that although a segment could be removed in this manner, the segment that was still attached was, due to its hanging condition, no longer part of the cover. The court however noted that the characteristic claimed in the patent was the ability to remove one segment while leaving the adjacent segment(s) still *attached*, hanging or not, as the Sundance patent recited in its claims. The court therefore found the combination to be obvious, observing that "[i]n combination with each other, the cover system of Cramaro [the 'tarpaulin cover system' patent] performs exactly the function that it performs independent of Hall [the 'flexible canopy structure' patent], and the removable cover sections of Hall perform exactly the function that they perform independent of Cramaro."[9]

The clearest case of a combination of known elements where each element does more than it did on its own is one in which the combined elements produce a synergistic effect. Synergism is often shown and argued in chemistry-related inventions where it can be demonstrated by test data. Inventions in agricultural chemistry are prime examples, since pest control and crop growth rates are easily obtained and compared. Bayer Cropscience's Patent No. 8,524,634,[10] for example, claims a method for reducing damage to seeds or plants by applying a combination of

[9] *Sundance, Inc., et al. v. DeMonte Fabricating Ltd. et al.*, 550 F.3d 1231 (Fed. Cir. 2009).

[10] "Seed Treatment With Combinations of Pyrethrins/Pyrethroids and Clothianidin," United States Patent No. 8,524,634; Asrar et al., inventors; Bayer Cropscience AG, assignee; issued September 3, 2013.

clothianidin and any member of a class of chemical compounds known as synthetic pyrethroids. Clothianidin and pyrethroids were already known at the time of the invention as insecticides for crop control, but they were nowhere disclosed as a combination. The patent application contained test data on corn plants infested with black cutworm, a common crop pest, in which the seeds of the corn plants were treated with clothianidin, tefluthrin (a synthetic pyrethroid), or both, at various application rates (grams of insecticide per kilogram of seed), all compared with an untreated control. The data included both stand reduction (percent loss of the corn plants) and percent control (of the cutworm) for each test and showed that both values were greater when the two insecticides were applied as a combination than when applied individually, and the comparisons included tests with the same total application rates. This showing established the combination as nonobvious and resulted in grant of the patent.

Synergism is difficult to show in nonchemical inventions, but the same result can occur when an inventor shows that individual elements when combined function differently than they did in the prior art. A case in point is Patent No. 7,713,133,[11] where the combination is a tennis court substrate and particles of waste glass ("cullet") of a specified size range adhering to the surface of the substrate. The prior art cited against the claims included an earlier patent disclosing cullet for various uses, including "as an aggregate or filler in roadway paving." This was combined with the inventor's own admission (in the patent application itself) that clay and claylike tennis courts made of a granular material were already known. The inventor's argument was that the cullet performed a different function on the surface of the substrate than it did as an aggregate or filler and that neither the known use of granulated clay for tennis courts nor the reference patent's description of the use of cullet as a filler would provide sufficient reason for one or ordinary skill in the art to place cullet on the surface of the paving material rather than to mix it in with the clay to form the bulk of the paving material. The examiner did not agree with the argument, but the Appeals Board did,[12] and the patent issued 9 months after the Board's decision.

7.1.2 "But Why Would a Munitions Manufacturer Go to a Horse Trainer (for the Missing Element)?"

Most likely it would not. When individual elements of a combination perform the same functions that they were known to perform individually, the combination will only be obvious if the "arts" in which the individual elements are originally found are "analogous." Patent courts and the USPTO warn that "analogous" is to be interpreted broadly in this context, but inventions can still be found where the sources of the elements were sufficiently removed from each other ("nonanalogous") that combining the elements was declared to be nonobvious.

---

[11] "Surface Composition for Clay-Like Athletic Fields," United States Patent No. 7,713,133; Wolf, inventor; issued May 11, 2010.

[12] *Ex parte* Ann Marie Wolf, Decision on Appeal, USPTO Board of Patent Appeals and Interferences, August 18, 2009.

The case of Anchor Holdings' Patent No. 8,224,669[13] is an example. The invention in this patent is a health-care management system for use by patients and their clinicians in the management of chronic diseases, with diabetes listed as a leading example. The system as expressed in the first (and broadest) claim of the patent has four components, the fourth having four subcomponents. The components (paraphrased) are as follows:

1. A database on a server computer, containing medical, education, exercise, and diet information relating to the management of a patient's disease

2. A patient component accessible by the patient and in communication with the database to allow the patient access to the information on the database

3. A clinician component accessible by a clinician and in communication with the database to allow the clinician access to the patient's information

4. An interactive education component between the patient and the clinician, containing:

   (a) A data store containing medical reading material from the database

   (b) A listing of reading material specifically selected by the clinician from the data store for the patient

   (c) A viewer by which the patient can view the selected reading material, the viewer indicating to the database when the material has been read

   (d) A clinician alerter for informing the clinician that the patient has read all the selected materials

The patent examiner rejected this claim over a combination of three earlier patents, one directed to a system for monitoring a chronic disease using a database of patient data entries, another directed to a therapeutic modification program including a compliance monitoring and feedback system, and the third directed to a "System and Method for Collecting, Processing, and Distributing Information to Promote Safe Driving." This third patent, which was intended for use by automobile insurance carriers, was cited for the component of reporting back whether instructional materials were read, which corresponded to subcomponent 4(d) of the Anchor Holdings claim.

Anchor Holdings appealed the rejection, and the Appeals Board[14] agreed with Anchor Holdings and reversed the rejection, holding the third reference to be non-analogous either to the first two or to the Anchor Holdings invention. The Board explained that "The analogous-art test requires a showing that a reference is either in the field of the applicant's endeavor or is reasonably pertinent to the problem with which the inventor was concerned in order to rely on that reference as a basis for rejection" and that the examiner had not shown either. The patent was granted 7 months later.

---

[13] "Chronic Disease Management System," United States Patent No. 8,224,669; Peterka et al., inventors; Anchor Holdings, Inc., assignee; issued July 17, 2012.

[14] *Ex parte* Bruce A. Peterka et al., Appeal No. 2010-007081, Decision on Appeal, USPTO Board of Patent Appeals and Interferences, December 15, 2011.

7.1.3 "But Nobody Knew What the Problem Was (Before I Came Along)!"

Even when the elements of a combination perform the same function that they were known to perform individually and are found in "analogous arts," the combination can still be nonobvious if the invention lies in the discovery of a problem or of the cause or source of a known problem. An obvious solution can thus lead to a patentable invention if the problem itself (or its cause or source) was not obvious.

This was indeed the case in a lawsuit involving two patents[15] on formulations for omeprazole, the active ingredient in Prilosec, a drug used to treat symptoms of gastroesophageal reflux disease (GERD), gastric and duodenal ulcers, and other conditions caused by excess stomach acid. Omeprazole was most effective when released in the small intestine. The environment in the small intestine is alkaline, but to reach the small intestine, the drug had to pass through the stomach where the environment is acidic. Prior to the inventions in these patents, it was known that omeprazole would degrade in acidic media and upon contact with moisture. This reduced the activity of the drug by causing it to degrade before it reached the small intestine. Enteric coatings containing acidic compounds were known to prevent much of this degradation, but the acidic compounds in these coatings themselves caused the drug to decompose, particularly in storage. To address this problem, alkaline-reacting compounds were added to the drug core to counteract the acidic compounds in the enteric coatings. The degradation and storage instability problems persisted nevertheless.

The inventors in the two patents speculated that the problem might lie in an interaction between the alkaline-reacting compounds in the core and the acidic enteric coating and decided to add a subcoating between the drug core and the enteric coating. They tried a number of substances for the subcoating, including water-soluble substances even though they feared that a water-soluble subcoating might dissolve in the water from the gastric juices that diffused through the enteric coating, leaving the water-sensitive omeprazole susceptible to degradation from the water. To their surprise, the tests revealed not only that the water-soluble subcoating improved the drug's storage stability, but it also improved the drug's resistance to the gastric acid. The patents thus claimed the combination of a core containing both the drug and the alkaline-reacting compound, a water-soluble (or water-dispersing) subcoating, and an enteric coating.

When the patent owner Aktiebolaget Hassle sued Apotex for patent infringement, Apotex responded with several defenses, including one of invalidity due to obviousness, based on a combination of several references. One of these references was a published European patent application that disclosed an omeprazole tablet that included a drug core containing omeprazole without an alkaline-reacting compound, the core covered by an enteric coating of cellulose acetate phthalate, one of the coating

---

[15] "New Pharmaceutical Preparation for Oral Use," United States Patent No. 4,786,505; Lovgren et al., inventors; Aktiebolaget Hassle, assignee; issued November 22, 1988; and "Pharmaceutical Formulations for Acid Labile Substances for Oral Use," United States Patent No. 4,853,230; Lovgren et al., inventors; Aktiebolaget Hassle, assignee; issued August 1, 1989. The two patents expressed the invention in different ways, but for the purposes of this discussion, both covered the same subject matter.

materials covered by the Aktiebolaget Hassle patents. The other references disclosed various drugs that had both an enteric coating and a subcoating, but did not disclose omeprazole or any drug that behaved in the same way in the stomach or intestine. Aktiebolaget prevailed against all of the defenses and the patent was upheld. To respond to the obviousness allegation, Aktiebolaget explained that the European application did not suggest any need to stabilize the omeprazole, nor did it provide any suggestion of a negative reaction between the enteric coating and the drug core. The inventors' discovery that the interaction was the cause of the instability therefore established the nonobviousness of the invention.[16]

7.1.4 "But They Said It Couldn't Be Done!"

Recall that the notion of a prior art reference "teaching away" from an invention was mentioned in the discussion of novelty in Chapter 6, where it was shown that the patent system does not consider "teaching away" a persuasive argument in favor of novelty. Thus, if an invention is fully disclosed (all of its elements are shown or stated in the same combination or arrangement) in a single document of prior art, the document anticipates the invention regardless of any negative statements that the document might make regarding the viability of the invention. In a nonobviousness determination, however, a "teaching away" can be to an inventor's benefit. When it is necessary to combine two or more documents (or sources in general) of prior art to show all the elements of the invention and one of the documents teaches away from the other or otherwise suggests a negative result when incorporating or substituting an element from the other, nonobviousness can be established by showing that the combination produces a positive rather than a negative result.

An example of a "teaching away" that worked to an inventor's advantage is the case of Crocs, Inc.'s Patent No. 6,993,858,[17] directed to ventilated footwear similar to "flip-flops" or sandals but with the added feature of a strap made of a foam material. The beneficial effect of using foam for the strap was that it provided friction between the strap and the base section of the shoe where the strap was attached, the base section including both the sole of the shoe and the upper, with the strap attached to either. Once the wearer adjusted the strap to a desired angle, friction held the strap at that angle and provided what the inventor later characterized as a passive restraint system, that is, a loose anatomical fit that lent support to the shoe without continuously pressing against the Achilles tendon. When Crocs' overseas competitors began importing shoes that met the patent claims, Crocs filed a complaint with the International Trade Commission (ITC), and one of the defenses raised by the importers was invalidity of the patent due to obviousness. Two sources of prior art were cited, the "Aqua Clog," which was a preexisting shoe containing a base section but no strap, and a patent to Aguerre (No. 6,237,249) that disclosed a sandal with a

---

[16] *In re* Omeprazole Patent Litigation, 536 F.3d 1361 (Fed. Cir. 2008).

[17] "Breathable Footwear Pieces," United States Patent No. 6,993,858; Seamans, inventor; Crocs, Inc., assignee; issued February 7, 2006.

heel strap attached to the sandal base section. The Aguerre patent described the heel strap as necessarily elastic, however, and stated a need to avoid friction between the heel strap and the remainder of the shoe, designing the connection between the strap and the base section as one that would allow the strap to "more freely rotate." The ITC nevertheless found this distinction unpersuasive and declared the Crocs patent invalid for obviousness over the combination of the Aqua Clog and the Aguerre patent. The appeals court reversed,[18] however, noting that by describing friction as a problem, the Aguerre reference "taught away" from the beneficial use of friction in the Crocs invention, and the Crocs patent was finally held to be valid.

Most inventions of course are not the result of successfully attempting something that the prior art says not to. There are other ways, however, in which an invention can be functional and yet contrary to a suggestion or "teaching" of the prior art and therefore nonobvious. This can occur, for example, when the modification needed to the prior art to arrive at the invention would make the prior art unsatisfactory or inoperable for its intended purpose.

Such was the result in Siemens Corporation's Patent Application No. 11/802,209,[19] which claims an apparatus for reducing particulate matter and NO_x emissions from a diesel engine by injecting a small amount of diesel fuel into the exhaust stream. A drawing from the application is shown as Figure 7.2.

The apparatus includes a fuel injector (30) with a solenoid valve (32), and to prevent damage to the valve from the hot combustion gas produced at the injection site, the apparatus includes a flow-through water jacket (34) surrounding the valve at a location that is "integral with and adjacent to" the solenoid coil. The prior art cited against the claims was a 1933 patent that described a fuel valve with an atomizing nozzle and a flow-through water cooling chamber. The valve was a needle valve shaped to seat against the inner end of the nozzle. The means by which the needle valve was actuated was not disclosed in the patent, but solenoid valves were well known by the time Siemens filed its application, and the examiner rejected the claims, reasoning that the placement of a solenoid coil in the reference valve was obvious. Siemens responded by pointing out that the cooling chamber in the 1933 patent was positioned and designed to surround and cool a "nozzle carrying member," that is, a block with an internal passage that was shaped to hold the atomizing nozzle and to "guide and receive" the needle portion of the valve. If one were to add a solenoid coil at a location where the cooling chamber would cool the portion of the valve that was "integral with and adjacent to" the coil, as the Siemens claim recited, the coil would interfere with the nozzle carrying member's ability to guide and receive the needle (in accordance with the description in the 1933 patent). The modification would thus make the structure in the 1933 patent inoperable for its intended purpose. This argument succeeded on appeal, and the Appeals Board[20] declared the Siemens invention nonobvious.

[18] *Crocs, Inc. v. International Trade Commission et al.*, 598 F.3d 1270 (Fed. Cir. 2004).

[19] "Automotive Diesel Exhaust Water Cooled HC Dosing," United States Pre-Grant Patent Application Publication No. US 2007/0290070 A1; Hornby, inventor; Siemens VDO Automotive Corporation, assignee; published December 20, 2007.

[20] *Ex parte* Michael J. Hornby, Decision on Appeal, USPTO Patent Trial and Appeal Board, August 30, 2013.

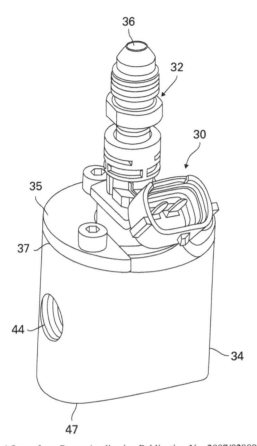

Figure 7.2   Selected figure from Patent Application Publication No. 2007/0290070.

The same result occurs when modification of the teachings of one reference by incorporating those of another reference would change the principle of operation of the first reference. Universal Laser Systems' Patent No. 8,599,898[21] is an example. A drawing from the patent is shown as Figure 7.3.

The invention in this patent is a slab laser with a resonator formed by two reflective surfaces (22, 31) positioned at a nonparallel angle to each other, with at least one of the surfaces having two or more reflective regions (32). The reflective regions together with the angle of the surfaces produce a modification of the phase distribution of the incident laser radiation propagating from the individual reflective regions, and the resulting beams are then incoherently[22] combined into a single

---

[21] "Slab Laser With Composite Resonator and Method of Producing High-Energy Laser Radiation," now granted as United States Patent No. 8,599,898; Sukhman et al., inventors; Universal Laser Systems, Inc., assignee; issued December 3, 2013.

[22] For readers who are not familiar with these terms, "coherent" in this context means "of a single wavelength," whereas "incoherent" means "of multiple wavelengths or spanning a range of wavelengths."

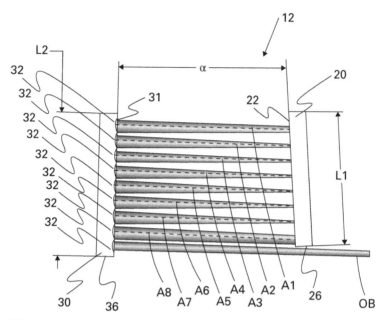

Figure 7.3   Selected figure from Patent No. 8,599,898.

output beam ("OB") of unusually high quality. When the application on which the patent was granted was examined, the claims were rejected over a combination of references of which two were the most prominent. One disclosed a slab laser with a "Talbot cavity" and with edge reflectors of a special construction, and the other disclosed a system for multiplexing the output beams from individual laser emitters in a common optical path by incoherently combining the beams and placing optical filters and partial reflectors within the path, one filter for each emitter and each filter being distinct from the others such that it and its reflector caused light from each emitter to reflect back to that emitter alone. The first reference had parallel rather than nonparallel reflective surfaces, and the second was cited for its inclusion of an optical wedge and therefore nonparallel reflective surfaces but used individual emitters to achieve the incoherent beams rather than creating them from a single beam by the configuration of the reflective surfaces. The rejection was reversed on appeal,[23] with the Appeals Board observing that the first reference was necessarily directed to the coherent combining of laser beams as a known characteristic of Talbot cavities, and the substitution of incoherent combining as in the second reference would alter this principle of operation of Talbot cavities. This led to the claims being declared nonobvious and the issuance of the Universal Laser Systems patent.

---

[23] *Ex parte* Yefim P. Sukhman et al., Decision on Appeal, USPTO Patent Trial and Appeal Board, May 10, 2013.

7.2 PURSUING THE UNPREDICTABLE

Many inventions fall within one or more of the categories listed in Section 7.1, but inventions by their nature are products of imagination and discovery and are seldom guided by a preestablished framework of arguments for nonobviousness. The best argument for many an invention is one that is unique to that invention. Nevertheless, if one were to identify a common theme, it would likely be unpredictability of success.

Unpredictability may be an elusive goal, but a class of innovations that clearly do not qualify are those that are matters of design choice or those that arise in response to market forces rather than being products of creative thought. Leapfrog Enterprises' U.S. Patent No. 5,813,861[24] was successfully challenged for merely being an upgrading of a preexisting device by applying new technology as a way of responding to market forces. The subject matter of the patent was an interactive learning device that teaches children how to read by associating letters and letter combinations with sounds. The device contains a keypad and a recess designed to hold specialized "books," plus a speaker and associated electronics. When the child touches a letter on the keypad or one of the words on an open page of the book, a reader on the device connects the touched letter on the keypad or the touched word on the book page with a voice processor circuit that pronounces the letter or the word through the speaker. In the book mode, the circuit first pronounces the entire word, then each phoneme of the word in sequence, and finally the entire word again. When Leapfrog sued its competitors Fisher-Price and Mattel for infringement of the patent, Fisher-Price/Mattel responded that the patent was invalid for obviousness over an earlier patent on a "Voice Puzzle Game" in combination with Texas Instruments' "Super Speak and Read" ("SSR") device and supplemented their defense with testimony from an expert witness at trial. The "Voice Puzzle Game" patent described a learning toy that contained a phonograph record, a speaker, an actuated electric motor to turn the record, and uniquely shaped puzzle pieces that fit into correspondingly shaped openings in the top of the toy's housing, each piece having a different letter imprinted on it. When a child depressed one of the puzzle pieces, the motor would cause the record to turn just enough to bring the phonograph needle to the location on the record that produced the sound associated with the puzzle piece. The SSR device was also a learning toy that produced sounds in response to a child pressing a letter, although with a different mode of operation. With the SSR device, the child would construct a word from individual letters or groups of letters and the device would then sound the word for the child to hear. The SSR device was introduced almost 20 years after the issuance of the "Voice Puzzle Game" patent and reflected the advances in technology that occurred in the interim by using a processor for associating the selected letter with the portion of a memory that produced the sound. Between the device in the "Voice Puzzle

[24] "Talking Phonics Interactive Learning Device," United States Patent No. 5,813,861; Wood, inventor; Leapfrog Enterprises, Inc.; assignee; issued September 29, 1998.

Game" patent and the SSR device, the only element lacking was a "reader" that sent the selected word to the processor, but the expert testimony at trial indicated that such a reader was common and well known at the time of the invention. The trial court agreed with Fisher-Price/Mattel that the combination of the references and the testimony made the invention obvious, and the appeals court confirmed,[25] observing that "[t]he combination is thus the adaptation of an old idea or invention [that of the device in the 'Voice Puzzle Game' patent] ... using newer technology [the processor and memory of the SSR device and the reader] that is commonly available and understood in the art." The patent was therefore declared invalid for obviousness.

If the upgrading had done something more than simply incorporating new technology into an old operation for an effect that the new technology was known to provide, the outcome may have been different. Similarly, if a new design such as a new arrangement, shape, or configuration does something more than provide an old construction with a new look, the new "design" may be nonobvious as well. Patent No. 7,467,680[26] of Ford Global Technologies, LLC, is an example. The subject matter of this patent is a hood of a car, and the invention lay in the understructure for the hood, that is, the underlying ribs that support and impart rigidity to the outer skin. The invention called for constructing the ribs in an "equilateral hexagonal pattern" as opposed to the more conventional manner that had a hub-and-spokes arrangement forming contiguous polygons, most of which were triangular but of different sizes. A drawing from the patent showing the ribs (12) and their hexagonal pattern is shown as Figure 7.4.

When the application for the patent was examined, it was rejected over two references, one of which showed the traditional hub-and-spokes arrangement with triangular cells of different sizes, while the other emphasized irregular rib patterns "designed to accommodate randomly spaced components [in the engine compartment under the hood] over which the structural panel is to be located." On appeal, the rejection was reversed,[27] the Board holding that "[w]e do not think that replacing the irregular ribs of [the two references] with the equilateral hexagonal ribs of the current invention is simply a 'design choice' which has no impact on the resultant hood. ... It is unpredictable whether the changes to an equilateral hexagonal pattern of reinforcing ribs for the hood would result in greater pedestrian survival at the cost of reduced survival of vehicle occupants, greater survival of both, or even reduced survival of both pedestrians and vehicle occupants." Ford's argument thus succeeded and the patent was soon granted.

---

[25] *Leapfrog Enterprises, Inc. v. Fisher-Price, Inc. and Mattel, Inc.*, 485 F.3d 1157 (Fed. Cir. 2007).

[26] "Motor Vehicle Hood With Pedestrian Protection," United States Patent No. 7,467,680; Mason, inventor; Ford Global Technologies, LLC, assignee; issued December 23, 2008. Although the invention was referred to as a "design" in the decision on appeal, the patent is a utility patent rather than a design patent (see Chapter 11 for the distinction).

[27] *Ex parte* David Edward Mason, Decision on Appeal, USPTO Board of Patent Appeals and Interferences, March 6, 2008.

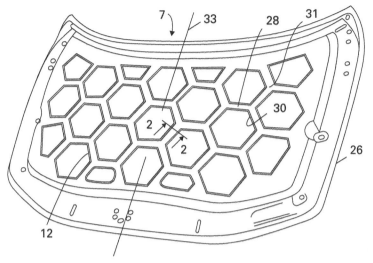

Figure 7.4   Selected figure from Patent No. 7,467,680.

7.2.1 Predictable Now but Unpredictable Then?

Any advance in the state of the art will make predictable what was once unpredict-able, and nonobviousness due to unpredictability therefore requires that the success of an invention be unpredictable at the time the invention was made. For this reason, a secondary reference (i.e., one used to modify or expand on a primary reference) can often be dismissed as prior art if its date is later than the filing date of the patent or patent application that it is cited against. References are often cited however for their statements of what was already known before the references themselves were pub-lished. Some of these references fail to state exactly when the information was known, however, or how much before their own publication dates. This raises the question of who, between the inventor and the examiner, has the responsibility of providing the date needed to establish that the information was indeed prior art.

Allergan's Patent No. 8,071,550[28] is directed to methods for treating uterine fibroids by the administration of a native, unmodified botulinum toxin. The application on which the patent issued went through two appeals within the USPTO before being declared allowable, the first appeal rejecting all of the claims. Among the rejected claims were those that included reducing distortion in the uterine cavity in addition to treating the fibroids. The rejection of these claims was based on a combination of ref-erences, one of which disclosed the ability of botulinum toxins to treat uterine fibroids by injection into the nervous system or by transdermal, peritoneal, subcutaneous, and various other means of administration, but did not disclose the local administration into a uterine fibroid or the reduction of any distortion in the uterine cavity. The reference that supplied this missing disclosure was a published article stating that an

[28] "Methods for Treating Uterine Disorders," United States Patent No. 8,071,550; Schiffman, inventor; Allergan, Inc.; assignee; issued December 6, 2011.

infertile woman might be treated by removing fibroids that distorted the woman's uterine cavity, thereby indicating that persons knowledgeable in gynecology already knew that distortion of the cavity could be reduced in this manner. This reference was published 7 months after the earliest effective filing date of the Allergan application, however, and Allergan responded to the rejection by stating that because of its late publication date, the reference did not qualify as prior art. The Board[29] found this argument inadequate, stating that "Appellant has not provided us with any evidence to convince us that those in the art first learned that fibroids distort the endometrial cavity less than seven months before [the reference] was published." Allergan ultimately amended the claims to overcome the rejection, but the Board's statement indicated that in this case at least it was the Appellant's (Allergan's) responsibility to find out when the information reported in the reference had originally become known.

In DMT Licensing's Patent No. 5,191,573,[30] the responsibility was placed on the examiner. This patent, which was applied for when music downloads were in their infancy, claimed a system and method for electronic sales and distribution of digital or audio signals that involved separate memories and the ability to transfer money and digital signals between those memories electronically. Included among the rejections that the application received in the USPTO was one that was based on a combination of references, the first reference disclosing a pay-per-view entertainment system, the second disclosing an electronic message unit that receives incoming calls and stores them in a memory, and the third suggesting (according to the examiner) the combination of the first two in view of its disclosure of a system for downloading a movie in digital format from an archive and allowing the user to store the downloaded movie locally and view it at any time. DMT objected to the rejection on the ground that the third reference was a patent whose filing date (and hence its effective date as prior art) was 6 months after the filing date of DMT's application. The examiner disagreed but was reversed on appeal,[31] the Appeals Board stating that "the Examiner has not shown where [the third reference] discusses the state of the art at or prior to the time of the current invention. Thus, relying on [the third reference] for motivation to combine [the first two] is improper," and the patent was ultimately granted.

Since the time differences in the two cases were virtually the same (7 months in the Allergan case and 6 in the DMT case), a possible explanation for the difference in the Board's approach might be the fact that the article cited in the Allergan case suggested that the information it disclosed was already known, while the patent cited in the DMT case did not. The difference may also however reflect the fact that the DMT invention arose at an earlier stage in its field of technology than the Allergan invention and hence the information was less likely to have been known much earlier than the reference's filing date. A third possibility for the difference in approach is

[29] *Ex parte* Rhett M. Schiffman, Decision on Appeal, USPTO Board of Patent Appeals and Interferences, September 8, 2008.

[30] "Method for Transmitting a Desired Digital Video or Audio Signal," United States Patent No. 5,191,573; Hair, inventor; Sightsound Technologies, LLC, most recent assignee, by assigned from DMT Licensing, LLC.

[31] *Ex parte* DMT Licensing, LLC, Decision on Appeal; USPTO Board of Patent Appeals and Interferences, September 4, 2009.

that the decisions in the two cases were rendered by different panels of administrative patent judges (each decision is rendered by a panel of three judges selected from a total of approximately one hundred).

7.2.2 Finding a Needle in a Haystack

An argument for nonobviousness that rarely succeeds is the proverbial "needle-in-a-haystack" argument, which goes something like this: "Sure, the missing piece that I added (or the modification I made) to arrive at my invention was in the prior art, but it was only one of a list of a huge number of possibilities and selecting it was like finding a needle in a haystack." The haystack argument tends to appear in inventions in the field of organic chemistry since complex organic compounds are often expressed generically and many of these genera cover untold quantities of compounds whose numbers are so large that they can only be estimated. When the argument succeeds, its success has little to do with whether the inventor actually had to screen an inordinately large number of candidates, than whether there was anything in the list that pointed the inventor in the right direction (or away from it).

The compounds in Takeda Chemical Industries' Patent No. 4,687,777[32] are thiazolidinediones, or TZDs, useful as a treatment for type II diabetes, and while the main claim recited a generic molecular formula, the formula contained only a single variable with a limited range limiting the genus to four compounds. One of these was pioglitazone, marketed in the United States under the trademark ACTOS®. The validity of the patent, which issued in 1987, was challenged by the generic drug manufacturer Alphapharm when Alphapharm filed an Abbreviated New Drug Application (**ANDA**) with the U.S. Food and Drug Administration (**FDA**) pursuant to the Hatch-Waxman Act for permission to market a generic version of pioglitazone. Alphapharm's challenge was that Takeda's claims were obvious over an earlier Takeda patent (issued in 1981) that listed TZDs as a large genus that included, but did not show, the four-compound genus or any of the four compounds themselves. One of the TZDs that the 1981 patent covered was a TZD that came very close to the four compounds, differing only by having a methyl group where each of the four compounds had an ethyl group. The 1987 patent presented comparative data comparing that methyl group-containing compound (which the 1987 patent referred to as "compound b") with one of the ethyl group compounds, showing that the ethyl group compound was superior. Takeda argued that its 1981 patent offered no suggestion of selecting "compound b" as a "lead compound" for comparison with the claimed compounds. (A "lead compound" is the compound in the prior art that would be most promising to modify when seeking to improve its activity.)

The court[33] noted that the genus in the 1981 patent covered "hundreds of millions" of TZD compounds, and although it included test results for nine compounds, including the compound later identified as "compound b," there was nothing

---

[32] "Thiazolidinedione Derivatives, Useful as Antidiabetic Agents," United States Patent No. 4,687,777; Meguro et al., inventors; Takeda Chemical Industries, Ltd., assignee; issued August 18, 1987.

[33] *Takeda Chemical Industries, Ltd., et al., v. Alphapharm, Pty., Ltd. et al.*; 492 F.3d 1350 (Fed. Cir. 2007).

in the 1981 patent to indicate that any of those nine compounds were the best performing antidiabetics. The court also considered a 1982 paper on antidiabetic agents that disclosed over a hundred TZD compounds, including compound b. The paper also identified three specific compounds as the most favorable, but these did not include compound b. A still further piece of prior art was included in the discussion—another Takeda patent that again covered over a million TZDs and even identified compound b as an "especially important" compound. The court nevertheless concluded that any favorable suggestion in this last patent regarding compound b was negated by the disclosure in the 1982 paper that acknowledged compound b but did not list it as one of the most favorable. Sorting through these disclosures, the court concluded that the three compounds identified in the paper as the "most favorable" would have been the most logical choices for a "lead compound," rather than compound b, and further noted that Takeda's comparative tests showed that compound b displayed toxicity and other adverse effects that would be incompatible with the long-term treatment needed for chronic diseases such as diabetes. Accordingly, given the huge number of TZDs in the prior art and the unfavorable results that Takeda had now shown for compound b, the prior art did not give any reason or motivation to select that compound as a starting point for the modification that would lead to pioglitazone or any of the four compounds claimed in the 1987 patent. Thus, the "needle-in-a-haystack" argument (although not expressed using those words) prevailed on the issue of what prior art compound should be considered a "lead compound," although it was bolstered by test data showing undesirable aspects of the lead compound in comparison to the claimed compounds.

In contrast, the needle-in-a-haystack argument did not succeed for Pfizer in its attempts to defend its Patent No. 4,879,303.[34] The patent is directed to the drug amlodipine, a long-acting calcium channel blocker useful as an anti-ischemic and antihypertensive agent. The invention lay in the use of the formulation of the drug as a benzene sulfonate (besylate) salt, which was found to improve the drug's chemical stability and reduce its stickiness in the tabletting process. Amlodipine besylate was marketed by Pfizer under the trademark NORVASC®, and an ANDA was filed by Apotex, Inc. for permission to market a generic version of the drug. In the ANDA, Apotex claimed that the Pfizer patent was invalid for obviousness based on a combination of Pfizer's original patent on amlodipine itself and references disclosing the use of pharmaceutically acceptable salts of drugs for various benefits. The besylate anion was listed among other pharmaceutically anions for the salts but not specifically for amlodipine. Pfizer's response to the challenge was that there were an unlimited number of pharmaceutically acceptable anions and that the length of time, expense, and difficulty involved in testing them to find the one that worked best should qualify the invention as nonobvious. One reference showed 53 anions, including besylate, that had been approved by the FDA, and of those, besylate was one of 23 that were used in less than 0.25% of the drugs. Other references showed that the besylate anion was known for its acid strength, solubility, and other favorable

---

[34] "Pharmaceutically Acceptable Salts," United States Patent No. 4,879,303; Davison et al., inventors; Pfizer Inc., assignee; issued November 7, 1989.

characteristics. The court held that 53 anions were not an unlimited number and that despite the time, money, and effort that would be needed to test all 53 to find the appropriate one for amlodipine, the testing was nevertheless routine. "These type [*sic*] of experiments used by Pfizer's scientists to verify the physicochemical characteristics of each salt are not equivalent to the trial and error procedures often employed to discover a new compound where the prior art gave no indication or suggestion to make the new compound nor a reasonable expectation of success"[35] (emphasis in original). For this reason, the claims were held to be invalid.

7.2.3 Unpredictability versus Optimization

Small changes in dimensions, proportions, or conditions can produce either small or big differences in properties, functionality, or utility, and many patent cases make a distinction between a difference in kind (one that is big enough to be patentable) and a difference in degree (one that is not). As is not uncommon in patent law, however, the dividing line is not evident, and the distinction between "in kind" and "in degree" is often more of a conclusion than a rule of law that would guide one toward reaching a conclusion. A somewhat clearer distinction is that between a beneficial effect that was unpredictable and one that is merely the result of an optimization of a known parameter. Here as well, the views of the USPTO and the courts may be best understood by examples.

Mere optimization was the conclusion for certain claims of Ecolab's Patent No. 6,113,963,[36] which is directed to the treatment of uncooked meat products to sanitize them by eliminating pathogens such as *E. coli* and salmonella. The claims in question recite the spraying of a certain combination of antimicrobials (a peroxycarboxylic acid with up to 12 carbon atoms and a carboxylic acid with up to 18 carbon atoms) onto the meat product at specified minimum concentrations and specified conditions of pressure, temperature, and contact time. When Ecolab sued FMC Corporation for patent infringement, one of FMC's arguments was that the claims were invalid for being obvious over one of FMC's own earlier patents. The FMC patent disclosed temperature and contact times that overlapped with those of the Ecolab claims, but said nothing about the spray pressure, which the Ecolab claims specified as "at least 50 psi." To address the spray pressure, FMC cited another prior art patent (one issued to Rhone-Poulenc, Inc.) that was likewise directed to killing salmonella and other bacteria on the surfaces of uncooked meat. Although the Rhone-Poulenc patent used a different antimicrobial than either of those listed in the Ecolab claims, it said that the antimicrobial was to be applied through spray nozzles at 20–150 psi "to vigorously wash the surface without damaging the meat." FMC's argument did not succeed at the trial court, but it did succeed at the appeals court,[37] which held that based on

[35] *Pfizer, Inc. v. Apotex, Inc.*, 480 F.3d 1348 (Fed. Cir. 2007).

[36] "Treatment of Meat Products," United States Patent No. 6,113,963; Gurzmann et al., inventors; Ecolab Inc., assignee; issued September 5, 2000.

[37] *Ecolab, Inc. v. FMC Corporation*, 569 F.3d 1335 (Fed. Cir. 2009).

the Rhone-Poulenc patent, "an ordinarily skilled artisan would have recognized the reasons for applying PAA [one of the peroxycarboxylic acids covered by Ecolab's claims] using high pressure and would have known how to do so." For this reason, the claims were declared invalid for obviousness.

Better success in rebutting an optimization argument was achieved by Agilent Technologies in its Patent Application No. 10/037,757.[38] The subject matter of the application was polynucleotide arrays, that is, two-dimensional arrays of extremely small spots on a flat surface, each spot containing DNA or RNA of a distinct nucleotide sequence. Arrays of this type are useful in screening large numbers of polynucleotides to find those that have certain types of activity or for testing a biological sample for diagnostic purposes to determine whether a particular substance was present in the sample by seeing which of the polynucleotides, if any, would bind to or react with the sample when the sample was applied to the entire array. As a material for the surface to which the spots were applied, glass offered certain benefits, but the screening and diagnostic procedures typically required the surface to be cut into small segments after the binding reactions had taken place, and this was difficult to do with glass. The invention solved this by using a base layer of plastic and applying the glass as a coating over the plastic. Among the claims in the application were two that limited the thickness of the glass layer to a range of 40–200 nm. The application explained that the thickness should be about one-fourth the wavelength of the light used for illumination of the array for digital reading but that it should also be thin enough to retain the flexibility of the layer, hence the 40–200 nm range.

Both claims were rejected as obvious over a combination of references, the lead reference being a patent that disclosed a probe carrier, that is, a microarray of DNA spots like the one in the Agilent patent application including a plastic base and a layer of glass over the plastic base. The reference patent made no mention of the thickness of the glass layer, but the examiner reasoned that "it would be obvious to one of ordinary skill in the art at the time the claimed invention was made to modify the glass thickness of [the reference] based on [the reference's] suggestion to do so … for the obvious benefits of optimizing the thickness to thereby optimize results." On appeal, this rejection was reversed,[39] with the explanation, "The examiner has not pointed to any evidence of record to support a conclusion that the thickness of the glass layer was recognized as a result-effective variable or that the size range recited in claims 7 and 18 would have been within the range of thicknesses that would have been considered obvious."

Reconciling these two cases suggests that if a particular type of adjustment is a known way to optimize, adjusting it will most likely be considered obvious, while adjustments that are not known to result in optimization and yet produce a benefit will have a better chance of earning a patent grant.

---

[38] "Chemical Arrays," United States Patent Application Publication No. US 2003/0108726 A1; Schembri et al., inventors; Agilent Technologies, Inc., assignee; published June 12, 2003.

[39] *In re* Carol T. Schembri et al., USPTO Board of Patent Appeals and Interferences Decision on Appeal, February 25, 2009. Despite the reversal of the rejection, Agilent did not advance the application to issuance as a patent.

7.3   IN HINDSIGHT (AND OTHER OBVIOUS OR NONOBVIOUS THOUGHTS)

Whether an invention is a combination of known elements, a substitution of equivalents or nonequivalents, a variation producing predictable or unpredictable results, or any other innovation, arising from an inspiration or otherwise, inventors are generally not pleased or satisfied when a patent examiner, a competitor, or a court of law dismisses their inventions as obvious. The preceding sections can provide a number of ways to respond to the first suggestion of obviousness, but others are used as well and frequently so. One of these is the hindsight argument, that is, that the invention should not be used as a roadmap to assemble components from the prior art regardless of their source or context or to make the same modifications that have already been made elsewhere and regardless of where. The difficulty with the hindsight argument is that it is little more than a conclusion, and a negative one, and rarely if ever succeeds on its own. It needs support from arguments that are positive in nature, those that focus on the assembled components or the modified product or process and point to some element of discovery in the quality or characteristics of the assembled or modified product. Another common response to a rejection or defense based on obviousness is to supply evidence of "secondary consideration," a loosely defined term that refers to such information as market data showing commercial success of the invention, a long felt but unresolved need that is finally resolved by the invention, the failure of others to achieve the result achieved by the invention, or even widespread copying of the invention once the application was published or the patent granted. These types of evidence are indeed successful at times, but they often fail. Commercial success, for example, will fail to persuade if it is attributable to features other than the novelty that distinguishes the invention over the prior art. Such features may be changes in market preferences or changes in the availability of materials that make the invention a more affordable alternative. The failure of others to achieve the same result may simply be due to a lack of funding to support their work in the subject area or their decision to focus on other research of more immediate interest or appeal to funding organizations. The perceived "copying" of the invention by competitors may actually be the result of competitors coming up with the invention on their own, which might indicate that the invention was more obvious than nonobvious. The most common reason that secondary consideration-type arguments fail is that they only apply to "close cases" of obviousness. Whether or not a case qualifies as "close" is a matter of degree, of course, but cases that are considered highly predictable are typically not considered close cases.

There is clearly no single definition of nonobviousness nor a single way of establishing it, but an overall approach might be to ask how the invention arose, why it was implemented, and what was learned from it. Somewhere in the answers to these questions, one can often find a basis for developing an argument in favor of nonobviousness.

Chapter 8

The View from the Infringer's Side: Challenging a Patent's Validity

Taking the Constitution of the United States at its word, the purpose of the U.S. patent system is "To promote the Progress of Science and the useful Arts," that is, to encourage technological advance by rewarding innovation and creativity, the reward consisting of "securing for limited Times to Authors and Inventors the exclusive Right to their respective Writings and Discoveries." For inventors, this means granting the inventor the power to exclude others from practicing the invention, despite requiring the inventor to provide to those who are excluded a set of detailed instructions on how to do precisely what they are excluded from doing. The logic may be debatable, but this has always ultimately been accepted as a general proposition, even when the power to exclude places those who are excluded at a competitive disadvantage. When the power to exclude has a destructive impact, the patent is often referred to as a blocking patent, a term that does not have a single definition. When loosely applied, it can simply mean a patent that prompts a commercial entity to redirect its business strategy. In extreme cases, however, it can mean a patent that prevents or inhibits commerce or business development in a major industrial or commercial sector and even one that prevents or inhibits technological advance itself. Whether its impact is large or small, a blocking patent can upset the plans of a start-up seeking to enter a market for its product or service or to develop or expand its customer base, an established company seeking to apply its expertise to a new product line or to expand into a different industrial sector, a company seeking to vertically integrate for cost advantage or quality control, or a company seeking to upgrade its product line to give a boost to its marketability or to meet current consumer demands. Knowledge of blocking patents can come from industry publications, from

First to File: Patents for Today's Scientist and Engineer, First Edition. M. Henry Heines.
© 2014 the American Institute of Chemical Engineers, Inc. Published 2014 by John Wiley & Sons, Inc.

information obtained at trade shows or through informal exchanges among researchers or business promoters, or through routine searching. Identifying blocking patents is one of the specific purposes of "due diligence," which in the patent context means the detailed investigation and verification of one's ability to control the competition through one's own patents while doing business unfettered by existing or potential infringement liabilities under the patents of others. Due diligence is a critical preliminary step in entering into joint ventures and other strategic alliances between companies, in mergers and acquisitions, and in fundraising including both public offerings and private placements.

The grant of a patent does not of course mean that the patent is valid, and scientists and engineers will occasionally encounter patents that are suspect on their face. This is often because of information that may have been more readily available or accessible to the person questioning the patent rather than to the patentee. An early commercial activity by someone other than the patentee that involves the invention in some way may be known only to the person involved or to a limited group of individuals rather than being common knowledge in the industry or information that can readily be found through conventional searching, even if no steps were taken to keep the activity confidential. A step or operating condition introduced in a plant or factory process without announcement to the industry or a need for regulatory approval is one example; others are a sale of a small number of units for market testing and an offer for sale that may not have been accepted: all can qualify as prior art damaging to a patent. Searching can also produce damaging prior art, particularly when the searching is more comprehensive than any searching that the patentee or the patent examiner would have done. The urge to challenge a patent's validity can also arise for reasons other than prior art: the breadth of its claims may be unreasonably large, the patent may misrepresent the invention's functionality, or the patent may violate fundamental principles of law by attempting to cover abstract ideas or basic scientific principles rather than practical applications. And of course, the desire to challenge may be prompted by a cease-and-desist letter from the patent owner, or even an informal threat by an employee or official of the patent owner.

8.1 DO YOU REALLY WANT TO GO TO COURT?

A patent's validity can certainly be challenged in a court of law, in an action instigated either by the patent owner or by one who is threatened by the patent owner, but the risk of loss and the costs involved in litigation can be daunting. In its 2013 Patent Litigation Study,[1] PricewaterhouseCoopers, LLP, reports that both the number of infringement suits filed and the number of ultrahigh damages awards were rising at the time of the study. Some of the increase in the number of suits filed is attributable

---

[1] C. Barry, R. Arad, L. Ansell, and E. Clark, "2013 Patent Litigation Study—Big cases made headlines, while patent cases proliferate," PricewaterhouseCoopers, LLP (2013). Available at http://www.pwc.com/en_us/us/forensic-services/publications/assets/2013-patent-litigation-study.pdf. Accessed May 26, 2014.

to the "antijoinder" provision of the AIA that requires patent owners who wish to sue multiple but independent infringers on the same patent to file a separate lawsuit against each infringer rather than joining them all in a single lawsuit. Nevertheless, the study found the increase to be beyond the amount attributable to the antijoinder provision. As for damages, the number of awards for patent infringement that exceeded $1 billion was the same in the year 2012 as in all years collectively prior to 2012. The median damages award for patent infringement has actually declined from 1995 to 2012 from slightly over $5 million to slightly under, but this is still a large number. The awards in cases decided by a jury however grew over the same time period, with medians rising from $6.9 to $12.2 million, suggesting that trying a case before a jury runs a particularly high risk. Costs are also entailed in the time between the filing of a lawsuit and the date that a verdict is rendered, since the uncertainty that prevails before a verdict is reached can make a party's financial projections difficult, as well as any business dealings and internal activities that may be affected by the outcome. The median of the time from the filing of the complaint to the trial date has been about 2.5 years since 2005 with little increase through 2012 except for a slow upward trend that is the result of the general rise in the number of cases going to trial each year.

Damages awards of course are only part of the potential cost of a lawsuit. Attorney fees will be incurred by both parties regardless of which party prevails, and the amount can vary with a wide range of factors, such as the degree of discovery, the number of pretrial motions filed, the number of witnesses called at trial, and all the unpredictable occurrences leading up to and during trial, including surprise testimony and changes in trial strategy. A 2011 survey by the American Intellectual Property Law Association[2] cites estimates of legal fees for different amounts at risk (potential damages awards): for up to $1 million at risk, the total legal fees, including disbursements, is estimated at $650,000; for $1–25 million at risk, the estimate is $2.5 million; and for $25 million or more at risk, the estimate is $5 million. In each case, more than half the legal fees will be incurred by the end of discovery.

What, therefore, are the alternatives to litigation? As we will see in the following sections, challenges to a patent or pending application can be made in the PTO in a variety of ways that differ, for example, in the window of time during which the challenge can be made, the grounds on which the challenge can be made, the requirements of and limitations on what submissions the challenger can make, the obligation of the PTO to respond to the challenge, the branch of the PTO that will hear the challenge, the extent of the challenger's continued involvement once the challenge has been filed, the finality of the result, and the expense. Whatever the form of the challenge, however, the challenge will be adversarial in nature, and to optimize one's chances for achieving the desired effect, whether it be a narrowing of the scope of

---

[2] Notably, the American Intellectual Property Law Association (AIPLA), letter from William G. Barber, President of the AIPLA to The Honorable Victoria A, Espinel, U.S. Intellectual Property Enforcement Coordinator of the Office of Management and Budget, Executive Branch of the United States Federal Government, RE: Request for Public Comments: Development of the Joint Strategic Plan on Intellectual Property Enforcement, August 10, 2012.

coverage, prevention of the issuance of a patent, or invalidation of an already issued patent, it is essential to understand what the patent covers and to know what claims one wishes to challenge.

8.2 SELECTING CLAIMS

A fundamental principle of patent law is that the scope of the patent owner's power to exclude is expressed in the claims, and while the specification (the text preceding the claims) often serves to illuminate or interpret the terms appearing in the claims, the focus for questions of coverage is on individual claims. In some patents, the specification and claims are not entirely consistent; the invention as described in the specification may be broader than the broadest claim, although this does not make the claims invalid. In any case, the claims themselves are of prime importance.

Claims commonly appear in sets that include both independent and dependent claims. An illustration appears below in which the term "widget" represents a manufactured article that is generic and hypothetical, used here to emphasize that the invention in these claims is fictitious.

Claims

1. A widget operating system comprising:

   (a) a widget enclosed by an electrically insulating housing and

   (b) first and second electrical contacts on an external surface of said housing, said electrical contacts being in electrical communication with said widget through electrical leads passing through said housing.

2. The widget operating system of claim 1 further comprising a power source external to said housing and connected to said widget through said housing.

3. The widget operating system of claim 2 further comprising a temperature detector within said housing.

4. The widget operating system of claim 3 further comprising a shut-off switch arranged to interrupt power from said power source to said widget when said temperature detector detects a temperature that exceeds a selected shut-off temperature.

5. The widget operating system of claim 4 wherein said shut-off switch comprises a manual control for setting said shut-off temperature.

6. The widget operating system of claim 3 further comprising an indicator light on said external surface of said housing connected to said temperature detector to emit visible light when said temperature detector detects a temperature that exceeds a selected shut-off temperature.

7. The widget operating system of claim 3 further comprising an audible alarm connected to said temperature detector to emit an audible signal when said temperature detector detects a temperature that exceeds a selected shut-off temperature.

8. The widget operating system of claim 1 wherein said housing comprises a base and an openable lid that when open exposes the interior of said housing to permit removal and replacement of said widget.

9. A widget operating system comprising:

   (a) a widget mounted to a mobile support base;

   (b) drive means for causing said support base to travel in two orthogonal directions; and

   (c) a manually operated controller governing said drive means.

10. The widget operating system of claim 9 further comprising side walls and a lid joined to said support base to form a housing enclosing said widget.

11. The widget operating system of claim 10 wherein said housing is electrically insulating.

12. The widget operating system of claim 11 further comprising first and second electrical contacts on an external surface of said housing, said electrical contacts being in electrical communication with said widget through electrical leads passing through said housing.

The independent claims in this claim set are claims 1 and 9, and the remaining claims are dependent, depending either directly from claim 1 or claim 9 or through intervening claims. The elements of claim 1 are a widget, a housing, and two electrical contacts with associated electrical leads. Operating systems that infringe claim 1 will be those that include these elements with any modifiers expressed in the claim, regardless of whether the operating systems also include other elements, such as additional contacts and/or leads, additional elements on or inside the housing, and particular features of the housing itself. Since the widget, the housing, and the two electrical contacts and associated leads as described are elements that an operating system must have if the system is to infringe this claim, these claim recitations are referred to as "limitations." Claim 2, by depending from claim 1, incorporates all the limitations of claim 1 and adds the limitation of a power source. As a result, claim 1 covers systems that contain a power source as well as those that do not, while claim 2 only covers those that contain a power source. Claim 2 is thus within the scope of, and yet narrower than, claim 1. Claim 3 adds the further limitation of a temperature detector to the limitations of claim 2 and is thereby within the scope of, and narrower than, both claims 1 and 2. Claims 4 and 5 add successive limitations, each claim being within the scope of, and successively narrower than, the claim that it refers to. The collective limitations of each of claims 2 through 5 are those of all lower-numbered claims, and each ultimately depends from claim 1; claim 3 through claim 2 as an intervening claim, claim 4 through claims 2 and 3 as intervening claims, etc. Claims 6 and 7 differ from those immediately preceding them by avoiding the incorporation of the limitations of claims 4 and 5 as intervening claims, and claim 8 depends only from claim 1 with no intervening claims. Claims 9 through 12 are a separate subset with no dependency from, or incorporation of the limitations of, any of claims 1 through 8, although certain claims of the 9 through 12 subset recite limitations that also appear in the 1 through 8 subset. All claims of both subsets differ in scope from each other, with no two claims within the entire set of 12 being identical in scope (this is a statutory requirement).

Before considering any of the various avenues for challenging the validity of the patent, one will want to determine whether any of the claims of the patent will actually be infringed by one's own product or process, and if so, which claims. This determination begins with the independent claims since they express the invention in its broadest scope. If a product or process falls outside all independent claims, the dependent claims will necessarily be avoided as well since the scope of each dependent claim is fully encompassed by that of the independent claim from which the dependent claim ultimately depends. Thus, if one wishes to market a widget operating system that contains a widget secured to a base of electrically insulating material that is open at the top rather than a housing that encloses the widget, the system will not infringe claim 1 nor therefore any of claims 2 through 8, regardless of whether the system includes an external power source, a temperature detector, a shutoff switch, or any of the other elements recited in claims 2 through 8. If the base is mobile with a drive means and a manually operated controller but the drive means only allows travel in one direction, the system will infringe neither claim 9 nor any of claims 10 through 12, regardless of whether the base contains side walls and a lid (claim 10), or whether the base is electrically insulating (claim 11), or whether it has two or more electrical contacts (claim 12). Note that the two independent claims overlap, that is, widget operating systems can readily be envisioned that infringe both claims 1 and 9, yet claims 1 and 9 must each be investigated for possible infringement. If one of both of these two independent claims cannot be dismissed in this manner, the dependent claims depending from the infringed independent claim(s) must then be studied to determine which of these will also be infringed. If, for example, the proposed operating system meets all limitations of claim 1 by enclosing the widget in an electrically insulating housing that contains a pair of electrical contacts on its surface joined to the widget through leads embedded in the housing and also contains an external power source, a temperature detector, and an audible alarm, plus a two-orthogonal-direction drive and a manually operated controller, the claims of concern will be claims 1, 2, 3, 7, 9, 10, 11, and 12.

In addition to using fictitious claims, the analysis itself is simplified. In an analysis of the claims of an actual patent or pending application, the entire record of the patent or application on file at the PTO must be reviewed, since statements or documents in the record may call into consideration additional legal doctrines that would extend the inquiry beyond the simple reading of the wording of each claim. The scope of a "means" clause, for example, such as the "drive means" of claim 9, is influenced by the description of the corresponding components in the specification, and products that contain elements that do not meet the literal scope of a claim limitation may still infringe under the "doctrine of equivalents." These and other factors involved in claim analysis find their basis in a body of court decisions and require the expertise of a qualified patent attorney.

Once the infringed claims are identified, the challenge, in the interest of both efficiency and completeness, can be focused on those claims that are infringed. For a challenge based on prior art, one will search for prior art in any form that anticipates (negates the novelty of) each infringed claim, or if one is already in possession of prior art, one will seek to satisfy oneself that the prior art anticipates each infringed claim. The anticipating prior art may be identical to the product or process that the challenger wishes to market or practice, but prior art that differs from the proposed product or process can still be effective to clear the product or process for liability-free marketing. In the widget claims example, any prior art disclosure of a widget operating system that falls within the scope of claim 7 will anticipate claims 1, 2, 3, and 7 of the first subset of claims, clearing from potential liability all widget operating systems that might fall within any of those claims, not just the operating systems of the prior art. Likewise, any prior art disclosure of a widget operating system that falls within the scope of claim 12 will anticipate all claims of the second subset, that is, each of claims 9, 10, 11, and 12, and clear from liability all widget operating systems falling within any of these claims. If both prior art disclosures are found or if a single disclosure meeting the limitations of both claims 7 and 12 is found, any operating system that encloses the widget in an electrically insulating housing that contains a pair of electrical contacts on its surface joined to the widget through leads embedded in the housing and also contains an external power source, a temperature detector, and an audible alarm, but does not contain a shutoff switch (claims 4 and 5), an indicator light (claim 6), or an openable lid (claim 8), will be clear of liability under the patent, that is, it will not fall within a single valid claim of the patent. The same analysis can be performed for any example of a widget operating system meeting one or more claims of the entire claim set.

Like the infringement analysis, this strategy for a prior art challenge is likewise simplified, since it addresses only anticipation and omits considerations of obviousness. As explained in Chapters 4 and 7, an investigation of obviousness is not necessary when anticipation is present. Nevertheless, the subtleties of each, like those of an infringement analysis, could fill volumes (and have), and the expertise of a patent attorney is essential for a full analysis. This chapter provides only an outline of these analyses.

For those seeking to challenge the validity of a patent, patent law offers several alternatives to litigation, each of which entails the consideration and resolution of the challenges in the PTO rather than in the courts. Some of these alternatives are new with the AIA, and others that were in existence before the enactment of the AIA are modified by the AIA. Each of these procedures offers a number of advantages over litigation, including faster resolution, lower cost to the challenger, and the expertise of the PTO, both in its understanding and implementation of the patent law and in the familiarity of the PTO officials with the technologies of the inventions. Although any invention will require some education of the PTO officials themselves, the PTO officials are accustomed to facing new technologies, and placing the decision-making process in the PTO avoids the need to educate a jury or a judge. The procedures have their limitations, however, and an understanding of each procedure will allow the challenger to select the procedure most appropriate for the particular challenge.

8.3 OPTIONS FOR CHALLENGE BEFORE THE PATENT IS GRANTED

If a flawed patent can be prevented from issuing or at least modified to eliminate or reduce any impact it may have on a competitor, the savings to the competitor in cost and time will be greatest, and the need to confront and respond to a patent owner may be avoided. Patent law provides two avenues for a "third party" (i.e., the challenger, the first party being the patent applicant and the second the examiner to whom the application has been assigned at the PTO or the PTO itself) to submit a challenge to a pending patent application and thereby influence the examination of the application. These avenues are (i) a submission of published material under Rule 290 ("Submissions by third parties in applications") and (ii) a protest under Rule 291 ("Protests by the public in pending applications") of the Rules of Practice in Patent Cases, which are part of Title 37 of the Code of Federal Regulations.[3] A protest, as its name implies, is somewhat more aggressive in nature than a mere third-party submission, but the two procedures overlap considerably in the opportunities that they offer to the third party as well as their requirements and limitations. The differences lie primarily in the bases for the challenge and the time limits for making the submissions.

Both a submission of documents under Rule 290 and a protest under Rule 291 can be based on printed publications, and as will be recalled from Chapter 2, the primary defining characteristic of a printed publication in the context of patent law is its dissemination to, or accessibility by, a significant portion of those segments of the public that are in the field of the invention's technology or interest rather than the physical form of the publication or the particular manner in which it is made accessible. Patents and published patent applications will in many cases be the printed publications that are most often used, but nonpatent publications can be relied on as well, such as technical articles, news media, doctoral dissertations, poster presentations and handouts, and the like, as outlined in Chapter 2. Qualifying publications under both rules are not however limited to those that qualify as prior art. Nonprior art publications can include those whose publication occurred too late to qualify as prior art, as well as litigation papers and court documents, except for those that are under protective or secrecy orders. Publications beyond prior art may, for example, cast doubts on the utility of the invention, indicate that test data in the application is misrepresented or misleading, or indicate an incorrect listing of inventors. Any published document that contains information having a potential bearing on the examination of the application can form the basis for a challenge under either rule.

Rule 291 protests extend beyond Rule 290 submissions in terms of information that is acceptable for submission: while both accept printed publications, Rule 291 protests can also be based on nonpublished information such as information revealing that a product or process within the scope of the invention was in public (commercial)

---

[3] Although the two rules differ in whom they name as the party making the challenge, neither the rules themselves nor the PTO's commentary appears to recognize a distinction between a "third party" and a member of the "public."

use or on sale. On-sale information may be in the form of an offer for sale, an accepted sales contract, or a completed sale, and this as well as public use information can be either confidential or nonconfidential, provided that they occurred within the appropriate time frame to qualify as prior art. Other examples are unpublished materials or other information indicating incorrect inventorship, an insufficient disclosure of the invention, or a misrepresentation of facts or data. Rule 291 thus offers more opportunities to protestors in terms of grounds for a challenge.

The PTO characterizes both Rule 290 submissions and Rule 291 protests as *"preissuance* submissions," but more accurately stated, they are *preallowance* submissions since both must be filed in the PTO before the application receives its Notice of Allowance, that is, the examiner's statement that all objections have been resolved and that the application is in condition for grant as a patent. This requirement is significant since issuance of the patent will only occur once the applicant pays the issue and publication fees, which the applicant is given 3 months from the date of the Notice of Allowance to do, and a few more months will then typically pass before actual issuance of the patent occurs. Certain applicants may even choose not to pay the necessary fees and thereby forfeit issuance, for any of a variety of procedural or strategic reasons, in which case a Rule 290 submission or a Rule 291 protest may no longer be necessary (although not prohibited). In any case, the period between allowance and issuance is not available to the third party for a submission under either Rule 290 or 291.

Between the two procedures, only the Rule 291 protest can be submitted up to the date of the Notice of Allowance, and if the protest is filed after the patent application is published, the protestor can only do so if the protestor first approaches the applicant and obtains the applicant's express written consent allowing the protestor to proceed with the protest. Without such a consent, the protestor must file the protest before the application is published, which generally occurs before the application receives its first **"Office Action"** (examination report from the examiner). A Rule 290 submission, by contrast, can be filed either before the first Office Action or 6 months after publication of the application, whichever is earlier, and in neither case requires the consent of the applicant. In the typical course of events, the publication occurs before the first Office Action, and the Notice of Allowance occurs last (although often with one or more intervening Office Actions and the applicant's responses to these Actions). If this is indeed the sequence, the person making Rule 290 submission will thus have 6 months more to file the submission than the Rule 291 protestor, unless the protestor obtains consent from the applicant, whereupon the protestor can file a protest up to the date of the Notice of Allowance.

Both procedures can be initiated without a filing fee, with certain limitations. A Rule 290 submission of printed publications can be made without a fee if the total number of publications submitted is three or less and the submission is the first and only such submission in that application by that third party. If the third party submits more than three publications, the fee is $180 for up to ten publications and an additional $180 for each additional 10. A Rule 291 protest can be made without a fee regardless of how many supporting documents are included, unless the protesting party (or the "real party in interest") has previously filed a protest of the same application. For second or subsequent protests by the same party, the fee is $140. Discounted rates are available for "small entities" and "micro entities" (these terms are defined in Appendix A), although a micro entity discount is not available for a Rule 290 submission. Assuming that the submitter or protestor uses a legal representative for either procedure, of course, the representative's fee will be the major portion of the cost. In either case, however, the fees are much lower than those for postissuance submissions, as will be seen in Section 8.4.

In both procedures, anyone can act as the third (submitting) party, including individuals, corporate entities, government agencies, and attorneys. Since a signature is required, the submitter (in either procedure) must be identified, but need not state whether he or she is the "**real party in interest**," that is, the party on whose initiative the submission or protest is made and who stands to benefit if the document submission or protest results in the patent not being issued or the claims being reduced in scope.

Table 8.1 Preissuance options available to a third party (challenger of the application)

	Submission of documents or information—Rule 290	Protest—Rule 291
Time limit for filing	Before the earlier of the examiner's first rejection of claim and the 6-month date after the pre-grant publication of application	Only up to the date of the **pre-grant publication** of application, unless done with the applicant's express written consent
Who can submit	Anyone; no need to identify the real party in interest	Anyone; no need to identify the real party in interest
What can be submitted	Patents, pre-grant publications of patent applications, and other printed publications, both qualifying as prior art and not qualifying as prior art	Patents, pre-grant publications of patent applications, and other printed publications, both qualifying as prior art and not qualifying as prior art, *plus* information other than printed publications that may be adverse to patentability, including information of commercial use, sale, or offer for sale
Continued participation of the submitter or protester after the submission or protest is made	None permitted	None permitted
Fee due to PTO	None for up to three items submitted; $180 (total) for 4–10	None unless not the first protest filed by the same protester or real party in interest in the same application

Both procedures require the party seeking to initiate the procedure to include a concise statement of the relevance of each document submitted. The statement must be very carefully prepared, since the submitter or protestor will have no further opportunity to augment the statement or to submit further documents or arguments. Also, once the initial submission is made, including all of the supporting documents required by the applicable rules, the submitter or protestor will receive no communication from the PTO other than a confirmation of receipt of the submitted papers, and this only if requested.

A summary of the differences between the two preissuance procedures is presented in Table 8.1.

8.4   OPTIONS FOR CHALLENGE AFTER THE PATENT IS GRANTED

A third party likewise has multiple options for challenging a patent after grant, and all of these options are either new with or modified by the AIA. The simplest is a "Citation of Prior Art and Written Statements" under Section 301 of the patent statute (Title 35 U.S. Code), which is similar to the Rule 290 "Submissions by third parties in applications" and expanded by the AIA from its earlier version. The more involved are a "Post-Grant Review" set forth in Section 321 of the statute and an "Inter Partes Review" set forth in Section 311 of the statute, both introduced by the AIA.

8.4.1   Citation of Prior Art and Written Statements

A "Citation of Prior Art and Written Statements" is a procedure by which a third party submits documents for inclusion in the official record of a granted patent. The submission is reviewed upon receipt at the PTO to determine whether it meets the requirements for entry into the record, but once entered, the documents are not reviewed by any examining authority within the PTO for any effect that they might have on the validity of the patent. Nevertheless, since the entire record of any granted patent, including all communications between the patent applicant and the examiner, is accessible to the public, the addition of these documents to the record means that any individual who later requests access to the record will see the documents.

This procedure is intended only for casting doubt on the validity of the patent or of individual claims of the patent over the prior art, and the documents submitted must fall under one of two categories. The first are printed publications, patent related or otherwise, and only those that would qualify as prior art, unlike the preissuance submissions discussed earlier, which can include nonpatent publications. The second category are documents filed by the patent owner in a proceeding before a federal court or in the PTO in connection with any patent application, in which the patent owner made a statement reflecting the scope of any claim of the patent for whose record the citations are now being submitted. The submitter can thus obtain documents (pleadings, deposition transcripts, trial transcripts, and the like) generated in lawsuits that the patent owner was involved in as well as the files of other patents that the patent owner has applied for, and look for statements in those documents showing that the patent owner's expressed view of what a particular claim covers is different from the actual wording of the claim or from an argument that the patent owner may have made in getting the patent granted. Such a statement may be damaging to the validity of a claim by effectively broadening the scope of the claim to bring it closer to, if not overlapping with, the prior art or by imposing an interpretation on a term in the claim in some manner to the detriment of the claim's validity. Whatever its category, each document must be accompanied by a detailed explanation of how it causes the claim to be invalid.

Citations under this procedure can be made at any time after the issue date of the patent and can be made by any individual, even the patent owner itself. The submitter must serve copies of all of the papers on the patent owner at the address of record, although the submitter can opt to have its name excluded from the patent file and to thereby remain anonymous.

Even though showing invalidity is its intent, a "Citation of Prior Art and Written Statements" does not make a patent invalid, since the PTO will take no action on the citation other than to place it in the patent's record. The patent owner may thus be able to devise its own strategy for defending the patent against the citations when the need to do so arises, such as in a later-filed lawsuit. Nevertheless, one advantage of the procedure is that it can be done without the need to pay a filing fee (although here again, of course, the submitter's legal representative will likely charge a fee). By contrast, both the Post-Grant Review and the Inter Partes Review have as their goal, and can result in, invalidation of the patent, and both require filing fees and rather hefty ones.

8.4.2 Post-Grant Review and Inter Partes Review

A "Post-Grant Review" and an "Inter Partes Review" are actual trials even though they are conducted within the PTO rather than in a court of law. Each trial is conducted before a panel of administrative patent judges (members of the Patent Trial and Appeal Board (PTAB)) rather than a jury or a judge in the federal court system. Like a conventional civil trial, however, each trial is a two-party adversarial proceeding in which either party can present testimony, file motions, and request oral hearings. Both the patent owner and the challenger have opportunities for arguments, and any document that either party submits in the proceeding must be served on the opposing party. The attorney fees will therefore be higher than those in the pre-grant procedures as well as the "Citation of Prior Art and Written Statements" procedure, both those before the patent is granted and those after. The filing fees (those charged by the PTO) for both the Post-Grant Review and the Inter Partes Review are bifurcated with one part due upon filing the request to initiate the proceeding and another upon granting of the request, and the amount of each fee varies with the number of claims being challenged. The filing fee for a request for a Post-Grant Review is $12,000 for up to 20 claims plus an additional $250 for each claim above 20, and the fee due upon grant of the request is an additional $18,000 for up to 15 claims plus

Table 8.2 Postissuance options available to a third party (challenger of the patent)

	Citation of documents or information—35 USC § 301	Post-Grant Review—35 USC § 321	Inter Partes Review—35 USC § 311
Time limit for filing	Any time after issue date	Within first 9 months after issue date	Any time after the 9-month date following the issue date
Who can submit	Anyone and can be filed anonymously	Anyone, but cannot be filed anonymously; and real party in interest must be identified	
What can be submitted	(i) Patents, pre-grant publications of patent applications, and other printed publications, but only those qualifying as prior art (ii) Documents from litigation or other patent applications, containing statements by same patent owner	Patents, pre-grant publications of patent applications, and other printed publications, both qualifying as prior art and not qualifying as prior art, *plus* information other than printed publications that may be adverse to patentability, including information of commercial use, sale, or offer for sale	Patents, pre-grant publications of patent applications, and other printed publications, but only those qualifying as prior art
Threshold for initiating the procedure	No requirements	Initial petition for the procedure must indicate that it is "more likely than not" that the petitioner will prevail	Initial petition for the procedure must indicate that there is a "reasonable likelihood" that the petitioner will prevail (a higher threshold)
Continued participation of the third party after the initial submission	None permitted	Extensive participation at same level as patent owner, once the initial request to initiate the procedure is granted	
Fees due to PTO	None	(i) Upon initial request: $12,000 for 20 claims or less (ii) Upon grant of request: $18,000 for 15 claims or less	(i) Upon initial request: $9,000 for 20 claims or less (ii) Upon grant of request: $14,000 for 15 claims or less
Result if successful and when	Simple entry into file; soon after submission accepted	Possible invalidation of challenged claims; within 1 year from beginning of review (following grant of request)	

$550 for each claim above 15. For an Inter Partes Review, the filing fee for the request is $9,000 for up to 20 claims plus an additional $200 for each claim above 20, and the fee due upon grant of the request is an additional $14,000 for up to 15 claims plus $400 for each claim above 15. No discounts are available in either proceeding for "small entities" or "micro entities."

The differences between these two proceedings are not reflected in their titles— indeed, both are "post-grant" and both are "inter partes." The differences instead lie in the windows of time in which they can be requested and the grounds for invalidity on which each can be based. A Post-Grant Review can only be filed *within* the first 9 months following the patent issue date, while an Inter Partes Review can only be filed *after* the 9 month date. As for grounds relied on, a Post-Grant Review has the wider scope, including not only patents and other printed publications that qualify as prior art but also nonpublished information, published material that is not prior art, and any information relevant to the utility, sufficiency, or accuracy of the disclosure in the patent. A Post-Grant Review can also be requested on the basis of "a novel or unsettled legal question that is important to other patents or patent applications." An Inter Partes Review, by contrast, is limited to patents and other printed publications and only those that qualify as prior art. There is also a difference in how the statute describes the thresholds that the third party must meet before the request for insti- tuting either type of review will be granted. The threshold for a successful request for a Post-Grant Review is that it be "more likely than not" that at least one claim is invalid, while the threshold for an Inter Partes Review is higher, specifically that there be "a reasonable likelihood that the petitioner [i.e., the requester] will prevail." Thus, a request for a Post-Grant Review can be granted if the odds indicated by the request are as low as just slightly greater than 50–50, while the grant of a request for an Inter Partes Review requires something more persuasive, although how much more is not articulated. Unlike submissions made pursuant to the "Citation of Prior Art and Written Statements," both the Post-Grant Review and the Inter Partes Review require identification of the real part in interest; neither can be filed anonymously or without identifying who is behind the request for the review.

In addition to avoiding the type of expense typically encountered in conducting or defending a civil trial in federal court, a further reason for the relatively low cost of Post-Grant Reviews and Inter Partes Reviews is that the regulations call for a final decision in both proceedings by the PTAB within 1 year of the institution of the pro- ceeding. This will limit the costs in terms of attorney fees and lost employee time, but it also offers the benefit of shortening the period of uncertainty in the outcome.

A summary of the differences between the three postissuance procedures is pre- sented in Table 8.2.

Chapter 9

Patent Eligibility: Pushing the Envelope on Subject Matter Appropriate for Patenting

The merits of an invention—its novelty, nonobviousness, and utility—determine its patentability, but the question of whether the subject matter of an invention is eligible for patenting asks whether the merits of the invention should be considered at all: Is this the type of invention that patents should be granted on, or is it an attempt to claim something that is too fundamental to commerce, industry, life, or nature to be made the exclusive property of an individual or a commercial entity, regardless of the ingenuity and resourcefulness behind its creation or discovery? Laws of nature, physical phenomena, and abstract ideas are recognized by the patent system as being patent ineligible for this reason, but as we shall see, the words themselves do not tell us where to draw the line. In many cases, these three categories are not even distinguishable from each other. The major problem that patent eligibility addresses however is that when the Information Age supplanted the Industrial Age, and as technology in all fields continues to advance to levels that were unimaginable centuries ago (when the first U.S. patent law was enacted), or even decades ago, the rules that determine what should or should not be patent eligible were, and are continually being tested, challenged, and often found to be no longer adequate.

Patent eligibility is not a question of breadth of coverage. Admittedly, over-breadth is a valid reason to reject a patent claim or to declare the claim invalid, but overbreadth is determined relative to the supporting description in the patent itself and in some cases the test data presented in the patent. Great breadth is not necessarily overbreadth, and a claim will not be held invalid simply because it is large in breadth. Nor is patent eligibility a question of the amount of time, effort, or financial resources invested by the inventor in arriving at the invention, the intellect of the inventor, or the circumstances by which the inventor arrived at the invention. Patent

First to File: Patents for Today's Scientist and Engineer, First Edition. M. Henry Heines.
© 2014 the American Institute of Chemical Engineers, Inc. Published 2014 by John Wiley & Sons, Inc.

eligibility is not even a question of the economic value of the invention. It is instead a question of public policy, rooted in balancing the interest of encouraging the innovator on the one hand and the undesirability of penalizing the industry as a whole on the other.

Official decisions by courts of law and the PTO repeatedly define patent eligibility by citing the three exceptions to eligibility—laws of nature, physical phenomena, and abstract ideas. The topic may be better understood however by examining the three fields of technology in which patent eligibility most often arises. These are (i) medical diagnostic methods, (ii) computer-implemented processes, and (iii) business methods.

9.1 MEDICAL DIAGNOSTIC METHODS

When viewed in the context of laws of nature, physical phenomena, and abstract ideas, medical diagnoses for particular diseases or classes of disease would seem to be relatively focused and limited in both purpose and scope. One might therefore expect the patenting of medical diagnoses to raise few if any of the public policy concerns expressed earlier. However, as medical research delves further and further into physiology, metabolism, and molecular biology to seek new approaches to prevention and cure, distinguishing diagnostic inventions from laws of nature becomes more difficult.

One case where a medical diagnostic method was held to be too much of a law of nature to be patent eligible was the case of *Mayo v. Prometheus*,[1] decided by the U.S. Supreme Court in 2012. There were two related patents in the case, and the Court focused on one of the two, Patent No. 6,355,623[2] as representative. The diagnostic method in this patent was for inflammatory bowel disease (IBD), an immune-mediated gastrointestinal disorder that encompasses a number of conditions including Crohn's disease and ulcerative colitis. Two drugs were known to be effective for treating IBD, 6-mercaptopurine (6-MP) and azathioprine (AZA), the latter actually being a "prodrug" that converts in the body to 6-MP. It was also known that 6-MP, either when administered directly to the patient or when formed in the patient's body from administered AZA, was itself converted by natural enzymes in the body to a number of metabolites, that is, molecules that are further modifications of, but still similar to, 6-MP. Both 6-MP and AZA were well known and in use for treating IBD, but they were not without problems. Certain patients failed to respond to the drugs, and others responded but suffered complications, including allergic reactions, neoplasia, and various opportunistic infections. What the inventions in the two patents provided was the

---

[1] *Mayo Collaborative Services, dba Mayo Medical Laboratories, et al. v. Prometheus Laboratories, Inc.*; 566 U.S. ___; 132 S. Ct. 1289 (2012).

[2] "Method of Treating IBD/Crohn's Disease and Related Conditions Wherein Drug Metabolite Levels in Host Blood Cells Determine Subsequent Dosage," United States Patent No. 6,355,623; Seidman et al., inventors, Hopital-Sainte-Justine, assignee, issued March 12, 2002. Prometheus Laboratories was the exclusive licensee of the patent and a manufacturer of a diagnostic test under the patent, and Mayo was Prometheus' customer until Mayo developed a test of its own embodying the same principles.

discovery that a particular metabolite, 6-thioguanine (6-TG), offered a direct indication of the efficacy of the drugs and also of whether the administered amount of either drug was too high and would therefore give rise to one or more of the complications. The claims of the patent recited the invention as a sequence of two steps, (i) administering the drug and then (ii) determining the level of 6-TG in a blood sample from the patient, and stated "wherein the level of 6-TG less than about 230 pmol per 8×10^8 red blood cells indicates a need to increase the amount of said drug subsequently administered to said subject and wherein the level of 6-TG greater than about 400 pmol per 8×10^8 red blood cells indicates a need to decrease the amount of said drug subsequently administered to said subject." The patent itself acknowledged that it was already known that 6-TG was a metabolite of 6-MP and that methods for determining the metabolite levels, including that of 6-TG, were likewise already known. The core of the invention was therefore the discovery of the correlation between the level of 6-TG and both the efficacy of the drug as a treatment for IBD and the toxicity of the drug as the cause of the undesirable side effects.

Prometheus Laboratories, Inc., was the exclusive licensee of the patents, and Mayo Medical Laboratories, Inc., was initially Prometheus' customer and then its competitor. When Prometheus sued Mayo for patent infringement, Mayo responded by arguing, among other defenses, that the patent was in effect claiming a law of nature and therefore not an eligible subject matter for a patent. The argument succeeded at the trial court but was reversed on appeal (by the Federal Circuit), and Mayo took the case up to the Supreme Court. The Supreme Court reversed the Federal Circuit and ruled in favor of Mayo, thereby changing the result back to that reached by the trial court, that is, that the invention was not patent eligible. In stating their decision, the Supreme Court justices observed that the relation between the amount of 6-TG in the patient's blood and both the efficacy of the drug and the likelihood that the drug would produce toxic side effects is a consequence of the way that the 6-MP is metabolized in the body and therefore a natural process or law. According to the Court, the "wherein" language in the claim stating the upper and lower limits and the indications of each simply informed doctors of this natural law "while trusting them to use those laws appropriately where they are relevant to their decisionmaking ... In so doing, they tie up the doctor's subsequent treatment decision whether that treatment does, or does not, change in light of the inference he has drawn using the correlations." The Court also observed that "the steps [of the claim] add nothing of significance to the natural laws themselves." Both patents were therefore held to be invalid for claiming patent-ineligible inventions.

An opposite result for a medical diagnostic method was reached within the PTO by the Board of Patent Appeals and Interferences (BPAI), soon after and acknowledging the *Mayo v. Prometheus* decision. The invention in Patent Application No. 10/984,320[3] was a method for obtaining a medical diagnostic image of a tissue or organ for such purposes as detecting abnormalities in organ or tissue perfusion. To determine

---

[3] "Ultrasonic Diagnostic Imaging System With Assisted Border Tracing," United States Patent Application No. 10/984,320, published on April 7, 2005, as Pre-Grant Publication No. US 2005/0075567; Skyba et al., inventors; ATL Ultrasound, assignee.

the exact location of an abnormality in a particular organ such as the heart, it was necessary not only to image the region but also to trace the borders of the organ and combine this information with the image. The invention provided a way of doing this by marking points on the border of the organ and then stretching or otherwise orienting a shape taken from a series of templates to connect the points. Although the emphasis in the patent application was on ultrasonic imaging of regions of the heart, the broadest claims did not specify either the particular type of image, the way it was generated, or the tissue or organ to be imaged. Claim 1 instead recited a three-step method for delineating the boundary, that is, (i) acquiring the image, (ii) manually marking three points of a boundary, and (iii) automatically fitting a predetermined border shape to the three points. The Examiner rejected the claims on several grounds, one of which was for being directed to a patent-ineligible process due to claiming an abstract idea. To arrive at this rejection, the Examiner applied the "**machine-or-transformation**" **test**, which is one of the recognized tests for patent eligibility[4] of a process and states that if the process involves the use of a machine or transforms something, it is patent eligible. Since the claims made no reference to a machine (or in this case an imaging apparatus), the Examiner took the view that the claimed methods "do not transform underlying subject matter (such as an article or materials) to a different state or thing."

The inventors took the patent application up on appeal to the BPAI, and after reviewing the claims, the BPAI disagreed with the Examiner and declared the claims to be patent eligible.[5] The BPAI reasoned that manually marking points on the image and fitting a border shape to the marked points amounted to a transformation of the image and that therefore the claimed method did indeed meet the "machine-or-transformation" test and was patent eligible for this reason. The application remained rejected on the other grounds, however, which included obviousness of the claims over the prior art, in this case another earlier-published patent application, and the application was ultimately abandoned. Nevertheless, the favorable decision on the issue of patent eligibility and the reasoning behind the decision were clearly stated.

Comparing this result to that reached in the *Mayo v. Prometheus* case, one notes that both cases reached one conclusion on patent eligibility and a different conclusion on patentability. In the Prometheus case, the claims were held to be nonobvious over the prior art (by the lower courts, a decision left undisturbed by the Supreme Court) and yet patent ineligible, while in Application No. 10/984,320, the claims were held to be patent eligible but obvious over the prior art. Thus, both inventions were ultimately deemed not patentable, but for different reasons. These cases demonstrate that patent eligibility and patentability over the prior art are two distinct issues and that meeting one does not necessarily mean that the other is met as well.

---

[4] This is not the only test, however, according to the Supreme Court in *Bilski et al. v. Kappos*, Under Secretary of Commerce for Intellectual Property and Director, Patent and Trademark Office, 130 S. Ct. 3218 (2010), in which a lower court rejecting a patent application for not meeting this test was challenged by the inventors arguing that it should not be the exclusive test and the Supreme Court agreed.

[5] *Ex parte Danny Skyba et al.*, Board of Patent Appeals and Interferences, Decision on Appeal, May 15, 2012.

It is also interesting to note that even though the inventions in the two cases were both medical diagnostic methods, the BPAI in the imaging method application framed its analysis in terms of whether the subject matter of the invention was an "abstract idea," while the Supreme Court in the Mayo decision used a "law of nature" analysis. The difference however is not significant, and the two terms are often grouped together. In any case, it is understandable in view of the Mayo decision to conclude that inventions in the medical diagnostics field will have a high chance of having their patent eligibility questioned, particularly if they reflect discoveries as to the biological or biomedical causes of diseases, conditions, and side effects. Nevertheless, the imaging method case, which succeeded and acknowledged Mayo, demonstrates that medical diagnostic methods can indeed be patent-eligible subject matter. The "machine-or-transformation" test may not be the overall standard, but it will most likely be a prominent test for inventions in medical diagnostics.

The field of gene identification and gene function is an area of medical diagnostics that has been the subject of great interest in both the medical and the legal communities. From the medical standpoint, newly acquired knowledge of the functions of certain genes, including mutations of those genes, has provided information useful in determining the likelihood that an individual will contract particular diseases, notably certain forms of cancer. Discoveries in gene function and genetic mutation are thus at the cutting edge of biomedical technology, offering the promise of major advances in the prevention, diagnosis, and treatment of disease. From the legal standpoint, the patenting of genes has been controversial, both on moral grounds and on the subject of patent eligibility. The eligibility question arises from the fact that genes and their mutations are naturally occurring. The identification of particular genes and their functions thus raises the question of whether they fall within the prohibition against patenting laws of nature.

Up until recently, the USPTO distinguished between genes in their native state, that is, as part of a much longer strand of DNA (i.e., chromosomes) in its natural form, and genes that are isolated from other genes and from the strand as a whole. The theory was that the isolation of genes required human intervention, that is, in cutting the genes out from their native chromosomes, and isolated genes were therefore not naturally occurring. This changed when the issue reached the Supreme Court in 2013,[6] regarding seven patents obtained by Myriad Genetics, Inc. These patents related to its discovery of the BRCA1 and BRCA2 genes by their locations in the human genome and their sequences, together with the discovery that certain mutations of these two genes are indicative of a significantly higher prevalence of breast cancer and ovarian cancer among women whose DNA contains these mutations. The Court focused on one of the seven patents, No. 5,747,282,[7] and specifically four claims of that patent, all directed to isolated genes of particular DNA sequences.

Isolation of a gene enables a clinician to study the gene sequence and thereby determine whether the mutation is present. Isolation can be achieved in two ways.

[6] *Association for Molecular Pathology et al., v. Myriad Genetics, Inc., et al.*, 133 S. Ct. 2103 (2013).

[7] "17Q-linked Breast and Ovarian Cancer Susceptibility Gene," United States Patent No. 5,747,282, Skolnick et al., inventors; Myriad Genetics, Inc., University of Utah Research Foundation, and The United States of America as represented by the Secretary of Health, co-assignees; issued May 5, 1998.

One is by extraction of the whole chromosome from a cell and then cleaving the chromosome at appropriate locations to remove segments other than the gene, leaving only the gene of interest. The other is by performing a series of chemical reactions in the laboratory to produce the gene synthetically, nucleotide by nucleotide (link by link of the chain), from a messenger RNA (mRNA) template. mRNA is itself naturally occurring and must be extracted from a living cell, but this synthetic method is actually more direct in certain respects than straight isolation. The gene produced by the synthetic method is referred to as complementary DNA (cDNA).

Despite the fact that the "isolated" DNA obtained by being cut from a chromosome and "isolated" DNA obtained synthetically (cDNA) are identical, the Court drew a distinction between the two, holding that cDNA did not violate the rule against patenting laws of nature, while the DNA cut from the chromosome did. Since the claims of the Myriad patent did not distinguish between the two, the claims were declared invalid as being drawn to patent-ineligible subject matter. Commentators have noted that this means that the many patents already granted to isolated DNA may all be invalid and that the reasoning behind the distinction raises numerous questions when applied to other species naturally occurring in the body as well as species derived from these naturally occurring species. This area of patent law will most likely continue to evolve. It is also interesting to note that the "machine-or-transformation" test did not play a role in the Court's decision, although "transformation" is certainly present in both routes to "isolated" DNA.

9.2 COMPUTER-IMPLEMENTED PROCESSES

The capabilities of a computer have opened up many fields to innovation and advancement, and due to the nature of software, computer-implemented inventions have given the "abstract idea" question a life of its own, separate from the concerns raised by processes occurring naturally in the human body or other "laws of nature" or "physical phenomena." A prominent question in the patenting of inventions involving software is: Does the inclusion of computer-implemented steps make a process patent eligible? Even the courts are not sure, and there are differences of opinion among the judiciary, including the judges of the Federal Circuit.

One case where the courts—first the trial court, then the Federal Circuit on appeal from the trial court, and finally the Supreme Court on appeal from the Federal Circuit—found that claims reciting a particular computer-implemented invention were patent ineligible for claiming an "abstract idea" was a case involving a series of four related patents owned by Alice Corporation Pty., Ltd.[8] These patents generally claimed a computerized trading platform for conducting financial transactions in which obligations between two parties are settled by an independent third party in

---

[8] "Method and Apparatus Relating to the Formulation and Trading of Risk Management Contracts," United States Patent No. 5,970,479, issued October 19, 1999; "Methods of Exchanging an Obligation," United States Patent No. 6,912,510, issued June 28, 2005; "Systems for Exchanging an Obligation," United States Patent No. 7,149,720, issued December 12, 2006; and "Systems and Computer Program Products for Exchanging an Obligation," United States Patent No. 7,725,375, issued May 25, 2010; all naming Shepherd as inventor, and all owned by Alice Corporation Pty. Ltd.

order to eliminate "counterparty" or "settlement" risk between the two parties, that is, the risk that one of the two parties will fail to meet its obligation and thereby cause the other party to suffer a loss such as an unreturned payment or an uncompensated performance. When Alice Corporation approached CLS Bank International with the patents, CLS objected and took the patents to court in a declaratory judgment action claiming that the subject matter of the patents was an abstract idea not eligible for patenting. The trial court agreed with CLS, and Alice Corporation took the case up on appeal to the Federal Circuit.[9]

Of the four patents, Patent No. 5,970,479 is representative, and the Federal Circuit summarized the leading method claim in this patent as follows.

> *The claim thus recites a method for facilitating a previously arranged exchange between two parties requiring the use of "shadow" records maintained by a third-party "supervisory institution." Briefly, the claimed process requires the supervisory institution to create shadow records for each party that mirror the parties' real-world accounts held at their respective "exchange institutions." At the start of each day, the supervisory institution updates its shadow records to reflect the value of the parties' respective accounts. Transactions are then referred to the supervisory institution for settlement throughout the day, and the supervisory institution responds to each in sequence by adjusting the shadow records and permitting only those transactions for which the parties' updated shadow records indicate sufficient resources to satisfy their mutual obligations. At the end of each day, the supervisory institution irrevocably instructs the exchange institutions to carry out the permitted transactions. Although [the claim] does not expressly recite any computer-based steps, the parties have agreed that the recited shadow records and transactions require computer implementation.*

The patents also included claims to computer-readable media and claims to data-processing systems whose components were designed to implement the method. Some of the judges at the Federal Circuit level made a distinction between the media and system claims on the one hand and the method claims on the other, while other judges did not. All however agreed that the method claims were not patent eligible, and the majority opinion stated as its reasoning that "the concept of reducing settlement risk by facilitating a trade through third-party intermediation is an abstract idea because it is a 'disembodied' concept." As for the computer implementation, the opinion stated that "simply appending generic computer functionality to lend speed or efficiency to the performance of an otherwise abstract concept does not meaningfully limit claim scope for purposes of patent eligibility." As for the media claims, these were likewise declared patent ineligible, with the reasoning that "we must look past drafting formalities and let the true substance of the claim guide our analysis." The system claims fared no better since they were considered to be "a computerized system configured to carry out a series of steps that mirror Alice's method claims … [and] taking a different approach to system claims would reward precisely the type of clever claim drafting that the Supreme Court has repeatedly instructed us to ignore."

At the Supreme Court level[10], the majority opinion of the Federal Circuit was confirmed, but unanimously, unlike the Federal Circuit. In regard to the method

---

[9] *CLS Bank International, et al., v. Alice Corporation Pty. Ltd.*, 717 F.3d 1269 (Fed. Cir. 2013).

[10] *Alice Corporation Pty. Ltd. v CLS Bank International et al.*, 573 U.S. ___, (U.S. Supreme Court, June 19, 2014), slip opinion.

claims, the Supreme Court held that "[u]sing a computer to create and maintain 'shadow' accounts amounts to electronic record keeping—one of the most basic functions of a computer ... The same is true with respect to the use of a computer to obtain data, adjust account balances, and issue automated instructions ... In short, each step does no more than require a generic computer to perform generic computer functions."[11] In regard to the other claims, the Court held that "Petitioner's claims to a computer system and a computer readable medium fail for substantially the same reasons."[12]

A month after Federal Circuit rendered its decision in the *CLS Bank International et al., v. Alice Corporation* case, the same court rendered a decision on another computer-implemented invention that also entailed a payment scheme, this time ruling in favor of patent eligibility. The patent that received this favorable ruling was Patent No. 7,346,545[13] of Ultramercial, Inc., whose invention was in the field of digitized music with concerns about the widespread practice of downloading MP3 music files from the Internet. As composers, authors, and performers have often complained, the downloading and sharing of these files has placed the music, as well as other copyrighted works such as literature, visual arts, and motion pictures, beyond the control of those who own the intellectual property rights to these works. Indeed, much of the downloading is done illegally, particularly by those who have Internet access but no credit or debit cards and by those attending colleges that provide high-speed Internet connections at no charge. In addition to the downloading problem, the invention also addresses the problems faced by advertisers whose banner ads on the Internet receive fewer and fewer click-throughs, as well as advertisers on television who have to deal with commercial-free cable channels, fast forwarding, and TV recorders that scan out commercials, even though the downloading problem and the advertising problems are unrelated.

Ultramercial solved both the downloading and the advertising problems with a single invention, which was an Internet and computer-based method for monetizing copyrighted products without any exchange of funds between the consumer and the copyright owner. In essence, the consumer receives a copyrighted product for free in exchange for viewing an advertisement that can be, but need not be, related to the copyrighted product. Claim 1 of the patent expresses the invention as "A method for distribution of products over the Internet via a facilitator" and lists the following steps: (i) receiving media products from a copyright holder, (ii) selecting an advertisement to be associated with each media product; (iii) providing the media products for sale on an Internet website; (iv) restricting general public access to the media products; (v) offering free access to the media products on the condition that the consumer view the advertising; (vi) receiving a request from the consumer to view the advertising; (vii) facilitating the display of advertising; (viii) for noninteractive advertisements, allowing the consumer access to the associated media product after the display; (ix) for interactive advertisements, allowing the consumer access to the product after the interaction; (x) recording the transaction; and (xi) receiving payment

---

[11] *Alice Corporation Pty. Ltd. v CLS Bank International et al.*, *supra*, slip opinion at p. 15.

[12] *Alice Corporation Pty. Ltd. v CLS Bank International et al.*, *supra*, slip opinion at p. 16.

[13] "Method and System for Payment of Intellectual Property Royalties by Interposed Sponsor on Behalf of Consumer Over a Telecommunications Network," United States Patent No. 7,346,545; Jones, inventor; Ultramercial, Inc., assignee; issued March 18, 2008.

from the advertiser for each display. The payment received from the advertiser is then used to pay royalties to the copyright owner.

Ultramercial sued three Internet companies for patent infringement, two of which were dropped from the case, leaving the third, WildTangent, Inc., as the sole defendant. WildTangent raised the defense of patent ineligibility of the invention on the "abstract idea" theory and succeeded at the trial court. Ultramercial took the case up on appeal to the Federal Circuit,[14] however, and succeeded in having the trial court's ruling overturned and the invention thus declared patent eligible. Once again, the Federal Circuit acknowledged that a method claim to an abstract idea does not become patent eligible by merely including in the claim a reference to a general-purpose computer, but it stated that tying the idea to a specific way of doing something with a computer or to a computer specially designed to implement the idea could render the claim patent eligible. The Court then applied the "machine-or-transformation" test, as in the cases cited earlier on inventions in the medical diagnostics field, and decided that the patent did more than simply claim the age-old idea that advertising can serve as currency; it involved an extensive computer interface in the various steps and substeps in the claim, enough to elevate it from a mere "abstract idea" to a patent-eligible invention.

Would the favorable decision in the Ultramercial case hold up in view of the Supreme Court decision in the CLS Bank case? Comparing the inventions in the two cases, the main difference in terms of the computer implementation aspect seems to be the fact that the central Ultramercial method claim recites a greater number of computer-implemented steps, that is, a more "extensive computer interface," than the central method claim in the CLS Bank case. The decision in the CLS Bank case however focuses on the individual computer implementation steps, asking whether each step is no more than a generic computer function, rather than on the extent of the computer interface. This casts some doubt on the viability of the Federal Circuit's decision in the Ultramercial case as well as the validity of numerous patents already granted on computer-implemented inventions, and for this reason the distinction between eligibility and non-eligibility in computer-implemented inventions may still be evolving.

9.3 BUSINESS METHODS

Many computer-implemented inventions are business methods, and the question of which aspect of any such invention is determinative of patent eligibility—the degree and nature of computer implementation or the underlying business process that the computer implements—may well depend on who is making the argument, the defender of the patent or its challenger, and on which statement among those made by the Federal Circuit or the Supreme Court will give the argument its strongest support. Business methods are often given special treatment, however, due in part to the fact that their patentability arose not from the emergence of a new field of technology such as computers but instead from an infringement suit relating to a patent

---

[14] *Ultramercial, Inc., and Ultramercial, LLC, v. Hulu, LLC and WildTangent, Inc.* (Fed. Cir. June 21, 2013).

on a data-processing system. The infringer invoked a long-standing policy in patent law that categorically declared business methods unpatentable, but the patent owner managed to convince the court to overturn the policy. The court's decision, *State Street Bank and Trust Company v. Signature Financial Group, Inc.*,[15] was published in 1998. Since then, the courts and the PTO have been tasked with the job of differentiating between business methods that are patent eligible and those that are patent ineligible for being abstract ideas.

As of late 2013, the most recent cases on the patent eligibility of business methods have been decided by the PTAB. One case, where the Board's conclusion was negative (i.e., the invention was declared patent ineligible), was that of Versata Development Group, Inc.'s Patent No. 6,553,350.[16] The invention in this patent was a method for determining a price of a product, and the invention arose from the need for field sales representatives to be able to conclude sales transactions quickly and completely in the field without having to send an inquiry to the home office and wait for the home office to sort through its records and send the sales rep the appropriate price for a particular customer. The invention involved arranging all of the company's purchasers into separate groups based on common characteristics such as the type of business each purchaser was engaged in, the frequency and volume of purchases made by each purchaser, etc., and doing the same for all of the company's products, that is, arranging them into separate groups on similar bases and then arranging the purchaser groups and the product groups into separate hierarchies, obtaining pricing information for the products and sorting it among the various groups, and drawing from the hierarchies to determine the price to charge a particular purchaser for a particular product. With the information organized in this manner, the sales rep could input the appropriate data for a proposed transaction through a laptop computer and receive back through the computer an appropriate price for the transaction.

Versata sued SAP America, Inc., for infringement of this patent and obtained a jury verdict of $351 million, which was affirmed on appeal by the Federal Circuit. SAP then petitioned the PTAB for review of the validity of the patent under a procedure newly introduced by the America Invents Act (AIA), entitled the **"Transitional Program for Covered Business Method Patents"** (discussed in the next section of this chapter). The petition was granted and the Board proceeded with the review. Upon completing its review, the Board concluded that the claims under review were not patent eligible and therefore not valid. The Board's reasoning was that the creation of organizational hierarchies for products and customers was a "disembodied concept" and "a basic building block of human ingenuity," that determining a price using organizational and product group hierarchies was akin to setting up management organizational charts, and that for these reasons the business method of the invention was an abstract idea. The Board also addressed the aspect of computer implementation, even though most of the claims under review did not recite

[15] *State Street Bank and Trust Company v. Signature Financial Group, Inc.*, 149 F.3d 1368 (Fed. Cir. 1998).

[16] "Method and Apparatus for Pricing Products in Multi-Level Product and Organizational Groups," United States Patent No. 6,553,350; Carter, inventor; Versata Development Group, Inc. (a change in name from "Trilogy Development Group, Inc." as listed on the patent); issued April 22, 2003.

particular computer-implemented steps. One of the claims recited a processor plus a memory that included computer program instructions and listed the instructions, but another claim simply recited "a method for determining the price of a product" and made no mention of programming or computer components. To bolster its argument for patent eligibility, Versata presented expert testimony to the effect that all practical implementations of the underlying process would be performed by computer, but under cross examination, the expert also admitted that the process could in theory be performed by pen and paper. Even if it were to read the claims as being limited to computer implementation, however, the Board held that implementation on a general-purpose computer (since no specialized computer was suggested by the invention) is not a meaningful limitation where the process has no substantial practical application other than by computer.

A case in which the Board found a business method to be patent eligible rather than ineligible was one that reached the Board directly from the Examiner corps, on appeal to the Board by the applicant from the Examiner's final rejection. The case was Application No. 12/418,430,[17] where the invention was addressed to online advertising and particularly to click-through ads hosted on websites with the advertiser paying the website publisher (the page provider) on a pay-per-click basis. When ads are placed on websites that are targeted to the end user of the merchandise promoted on the ad, more click-throughs occur, which translates to more business for the advertiser, more advertising revenue for the page provider, and more revenue for the ad broker who serves as a middleman between the advertisers and the page providers. The typical ad broker however has no way of knowing what websites the end user typically visits and thereby lacks the ability to profile end users and help with the targeting. The invention solves this problem by monitoring the online traffic associated with a particular end user and using the information thus obtained to develop a behavioral profile of the end user, which could then be offered to the advertiser, enabling the advertiser to place its ads on the web pages that are most likely to be visited by the end users that the advertiser seeks to target.

The claims that the Examiner had rejected as being patent ineligible included claims to a computer-readable storage medium as well as system claims and method claims. The rejection was reversed by the Board for all of these claims, but of particular note to the question of patent eligibility of business methods *per se* are the method claims since they make no reference to computer implementation or even to computer components. One claim simply recited "A method, comprising: determining a behavioral profile of an end user from monitored packet traffic initiated by the end user, the behavioral profile being determined by a controller operated by a first entity; and providing one or more systems access to the behavioral profile to identify an advertisement, wherein the one or more systems are operated by a second entity different from the first entity."

The Board explained its reversal of the Examiner by stating that the determining step in this claim explicitly required the use of a controller, thereby tying the claim to

---

[17] "Method and Apparatus for Presenting Advertisements," Application No. 12/418,430; Ou et al., inventors; SBC Knowledge Ventures, L.P., assignee; published as United States Patent Application Publication No. US 2009/0198569 A1, publication date August 6, 2009.

a particular machine. For this reason, according to the Board, the claim recited something more than purely mental steps, that is, steps that could be performed in the human mind, and therefore, the claim was not merely an abstract idea or concept. Here again, as in the cases discussed in the preceding sections of this chapter, the "machine-or-transformation" test saved the claim, even though the Board in this case as well made it a point to state that it was "mindful that the machine-or-transformation test is not determinative of whether an invention is a patent-eligible process."

9.4 THE AIA'S NEW PROCEDURE FOR CHALLENGING BUSINESS METHOD PATENTS

The cases in the preceding section seem to apply logical and straightforward reasoning, but the distinction that they make between patent eligible and patent inel-igible is a fine line, and many business method claims leave much room for argument on both sides, that is, both for and against patent eligibility. In fact, during the time period between 1998, when the State Street Bank decision opened the floodgates for business method patents, and the dates of the ensuing court decisions where much commentary was made and standards were evolved that distinguished patentable business methods from abstract ideas, many business method patents were applied for and granted without the benefit of those standards. In the Congressional hearings leading to the enactment of the AIA, the view was expressed that because of the relative lateness of the standards, a large number of business method patents that were granted may no longer be valid.

At the same time, concern has been mounting over the proliferation of patent "**trolls**," also known as "**nonpracticing entities**" (**NPEs**) or "**patent assertion entities**" (**PAEs**), and the number and nature of lawsuits that these entities have filed. Whether expressed in the pejorative or otherwise, the term refers to persons or com-panies that acquire patent rights from individual inventors or other small companies without themselves manufacturing the products or supplying the services covered by the patents (or ever intending to) and use the patents to coerce other companies, and often large numbers of them, into paying licensing fees to avoid the high costs of lit-igation, despite serious questions of the validity of the patents, actual infringement, or both. Although views on the subject differ, patent trolls have been accused of inhibiting innovation and technical advances and of clogging the court system with lawsuits of questionable merit.

The issue of questionable business method patents and the concern over patent trolls are not directly related, since trolls have been thought to be responsible for coercion on other types of inventions as well, notably software patents. Nevertheless, both the proliferation of questionable business method patents and the tactics used by trolls are considered to have led to the enactment of the section of the AIA bearing the title "Transitional Program for Covered Business Method Patents." This program, commonly known by its acronym the "**CBM** Program," provides challengers of business method patents with the opportunity to have the patents reviewed by the PTAB rather than a court of law, in an expedited procedure that is both less expensive and more quickly resolved than a lawsuit.

The terms "transitional" and "covered" in the title of the Program both limit the scope of the Program. The Program is "transitional" since it will expire on September 15, 2020 (unless extended by Congress); thus, any petition for a review under the Program must be filed by that date, although those pending as of that date will continue through to the completion of the review. The "covered" business method patents that the Program is directed to are those that claim "a method or corresponding apparatus for performing data processing or other operations used in the practice, administration, or management of a financial product or service" and does not include "technological inventions."[18] There are indeed many types of businesses that do not provide financial products or services and many types of business methods that are not related to such products or services, and accordingly, these cannot be challenged under this Program. The scope of "covered" business methods could be expanded by Congress as well, but as of the writing of this book, the scope remains that defined by the AIA.

According to the PTO regulations that govern how the PTO administers the Program, the Program follows the procedures for Post-Grant Reviews. Thus, one seeking to challenge a business method patent under the Program first files a petition to institute the review, and the review will be instituted if the petitioner can show that the challenge is "more likely than not" to succeed. Also, like a Post-Grant Review, a review once instituted under the CBM Program is a trial designed to reach a final decision within 1 year of the institution of the proceeding. The fees are also the same as those for a Post-Grant Review: a petition costs $12,000 for up to 20 claims plus an additional $250 for each claim above 20, and institution of the review when the petition is granted costs $18,000 for up to 51 claims plus $550 for each claim above 15. These once again are fees payable to the PTO and do not include attorney fees, which will typically be higher.

Differences between the CBM Program and other Post-Grant Reviews lie in who can petition to initiate the proceeding, which patents can be challenged, what grounds for challenging the patents can be made, and what types of prior art can be cited. Those seeking a review under the CBM Program are limited to individuals or entities that have already been sued for infringement or have received an actual threat of an infringement lawsuit. Thus, the scope of permissible challengers is more limited than those that can apply under the Post-Grant Review procedure. On the other hand, the scope of the review is broader in certain aspects than those for a Post-Grant Review. In particular, any patent, regardless of its issue date, can be the subject of the petition, except during the first 9 months following its issue date, and the validity of the patent can be challenged on any basis, rather than just prior art. Validity can even be challenged on the basis of "a novel or unsettled legal question," which is left undefined but opens up the procedure for use in applying further developments that may arise in the law of patent coverage for business method patents that cover financial products or services.

A summary of the most prominent features of the CBM Program is presented in Table 9.1, which can be compared to Table 8.2 of Chapter 8.

---

[18] Leahy-Smith America Invents Act, Section 18(d)(1).

Table 9.1 Reviews of business method patents under the Transitional Program for Covered Business Method Patents

Subject matter of claims that can be challenged	Method or corresponding apparatus for performing data processing or other operations used in the practice, administration, or management of a financial product or service, not including technological inventions
Time limit for filing petition	Any time following 9 months after the issue date, but petition must be filed before September 15, 2020
Who can file petition	Anyone who has already been sued for infringement of the patent or who has received an actual threat of a lawsuit; cannot be filed anonymously; real party in interest must be identified
What can be submitted to support the challenge	Patents, pre-grant publications of patent applications, and other printed publications qualifying as prior art
Grounds for challenge	Any ground of validity, including patent ineligibility
Threshold for instituting the review	Petition must indicate that it is "more likely than not" that the patent will be declared invalid
Continued participation of challenger	Full participation at the same level as patent owner once the petition for institution of the proceeding is granted
Fee due to the PTO	With petition: $12,000 for challenging 20 claims or less Upon grant of petition: $18,000 for 15 claims or less
Results if successful and when	Invalidation of challenged claims, within 1 year from the beginning of the review

9.5 CONCLUSION: A RULE FOR PATENT ELIGIBILITY? OR A CASE OF "I'LL KNOW IT WHEN I SEE IT"?

As the cases in this chapter demonstrate, the patent courts agree that laws of nature, physical phenomena, and abstract ideas do not qualify as patent-eligible subject matter. The Supreme Court says that "[t]he concepts covered by these exceptions are 'part of the storehouse of knowledge of all men … free to all men and reserved exclusively to none'."[19] The Supreme Court also acknowledges however that "[a]ll inventions at some level embody, use, reflect, rest upon, or apply laws of nature, physical phenomena, or abstract ideas,"[20] and courts have struggled to find ways to accommodate these seemingly contradictory notions. Of the three "exceptions," "abstract ideas" may be the most difficult to define, but the explanations that judges have given in identifying abstract-idea-containing yet patent-eligible inventions typify the difficulties in all three. The opinion in the CLS case, for example, states that what makes an invention that contains an abstract idea eligible is "whether it contains additional substantive limitations that narrow, confine, or otherwise tie down the

[19] *Bilski et al. v. Kappos, supra,* slip opinion p. 5, quoting *Funk Brothers Seed Co. v. Kalo Inoculant Co.,* 333 U.S. 127, 130 (1948).

[20] *Mayo v. Prometheus, supra,* slip opinion p. 2.

claim so that, in practical terms, it does not cover the full abstract idea itself."[21] The same opinion also states however that "[l]imitations that represent a human contribution but are merely tangential, routine, well-understood, or conventional, or in practice fail to narrow the claims relative to the fundamental principle therein, cannot confer patent eligibility."[22] To add to the confusion, the Supreme Court in the Mayo case warns that "the prohibition against patenting abstract ideas 'cannot be circumvented by attempting to limit the use of the formula to a particular technological environment' or adding 'insignificant postsolution activity'."[23]

The inquisitive inventor with more than a rudimentary awareness of patent issues will seek not only to innovate but also to direct her/his innovation in such a way that it will be patent eligible. For those with a technical mindset, the temptation is to look for "bright-line rules" or clear guidelines for patent eligibility. Unfortunately, the courts are of no help, responding that "[b]right-line rules may be simple to apply, but they are often impractical and counterproductive when applied to §101 [patent eligibility]. Such rules risk becoming outdated on the face of continuing advances in technology ..."[24] Even when general rules can be envisioned, we are told that "patent law's general rules must govern inventive activity in many different fields of human endeavor, with the result that *the practical effects of rules that reflect a general effort to balance these considerations* [i.e., the goal of patents in providing the incentive to create, invent, and discover vs. the risk of overly broad patents that impede the flow of information that might otherwise spur invention] *differ from one field to another*."[25]

In fact, even at the Federal Circuit and the Supreme Court, the two courts that are consulted by all trial courts and the PTO for legal precedents in patent law, it is not uncommon for either court to reach a decision for a particular invention but to decline to state a governing rule that can be applied to other inventions that have not yet come up. In the Ultramercial case, for example, the Federal Circuit, after stating its conclusion that the invention involved a sufficient amount of programming features to make it patent eligible, said:

> *Having said that, this court* does not define the level of programming complexity required *before a computer-implemented method can be patent-eligible. Nor does this court hold that use of an Internet website to practice such a method is either necessary or sufficient in every case to satisfy §101. This court simply holds the claims* in this case *to be patent-eligible ....*[26]

The confusion was even more explicit in the CLS case (discussed in Section 9.2) at least at the Federal Circuit level. The ten judges of the Federal Circuit divided themselves into five groups, each group expressing a separate opinion. The published

---

[21] *CLS Bank International v. Alice Corporation, supra,* slip opinion pp. 18–19.

[22] *CLS Bank International v. Alice Corporation, supra,* slip opinion p. 20.

[23] *Mayo v. Prometheus, supra,* slip opinion p. 4, quoting *Nilski v. Kappos* and *Diamond v. Diehr,* 450 U.S., 175 (1981).

[24] *CLS Bank International v. Alice Corporation, supra,* slip opinion p. 17.

[25] *Mayo v. Prometheus, supra,* slip opinion p. 23 (emphasis added).

[26] *Ultramercial v. Hulu, supra,* slip opinion p. 30 (emphasis added).

decision includes not only the opinion of the court as a whole (the "Per Curiam Judgment") but also those of each of the five groups. In a footnote to one of the opinions, the chief judge stated:

> *No portion of any opinion issued today other than our Per Curiam Judgment garners a majority. The court is evenly split on the patentability of the system claims.* Although a majority of the judges on the court agree that the method claims do not recite patentable subject matter, no majority of those judges agrees as to the legal rationale for that conclusion. *Accordingly, though much is published today discussing the proper approach to the patent eligibility inquiry, nothing said today beyond our judgment has the weight of precedent.*[27]

When the case was decided by the Supreme Court, on the other hand, the Court expressed no internal disagreement about the rationale for its decision nor did it seem to have any misgivings about the decision's value as precedent. As to what type of computer-implemented invention the Court would consider to be patent eligible, however, the Court provided only the barest of outlines, notably that the method claims for example would have to "purport to improve the function of the computer itself" or to provide "an improvement in any other technology or technical field."[28] While the Court did not need to go into further detail to render its decision on the invention before it, the intuitive inventor will note that if the invention is indeed an improvement in the function of the computer itself, the invention would be one relating to an innovation in basic software, firmware, or hardware rather than the technology or method that the computer is being used to implement, and would thereby shift and possibly obscure the purpose of the invention and its scope of applicability. Also, is not computer implementation "an improvement in some other (i.e., the underlying) technology or technical field"? For an invention that does entail an improvement in the underlying technology, the inclusion of computer implementation in the claim would only be appropriate if that were the only means by which that improvement could be achieved. The distinction may thus be between inventions where computer implementation makes possible the incorporation of an innovation in the underlying method (such inventions therefore being patent eligible) and inventions where the innovation is the computer implementation itself (such inventions therefore failing to be patent eligible). In any event, the distinction will be a challenge for both the legal community and the courts.

Indeed, in Bilski v. Kappos, referenced above and in numerous opinions by trial courts and the PTO, where the invention in question was a business method, the Supreme Court explicitly expressed a reluctance to state a general rule of law. The Bilski case was a patent application on hedging in the energy market, that is, the method by which buyers and sellers of commodities in the energy market can protect, or hedge, against the risk of price changes. The application was rejected by the PTO for failing to meet the "machine-or-transformation test," and the rejection

---

[27] *CLS Bank International v. Alice Corporation, supra*, Rader et al., opinion, concurring-in-part and dissenting-in-part, pp. 1–2, footnote 1 (emphasis added).

[28] *Alice Corporation Pty. Ltd. v CLS Bank International et al., supra*, slip opinion at p. 15.

was affirmed by the Federal Circuit. The inventors then appealed to the Supreme Court, acknowledging that the invention neither involved a machine nor transformed anything, but arguing that this test should not be the exclusive test for patent eligibility. The Supreme Court agreed but still found the invention to be patent ineligible as an abstract idea. The Court was clear in its statement that the "machine-or-transformation test" was not the exclusive test for patent eligibility, but the Court did not replace the "machine-or-transformation test" with any other overarching test. The Court instead stated the following:

> The Court, therefore, need not define further what constitutes a patentable "process," ... In disapproving an exclusive machine-or-transformation test, we by no means foreclose the Federal Circuit's development of other limiting criteria that further the purposes of the Patent Act and are not inconsistent with its text.[29]

As shown in the business method cases addressed in this chapter, the "machine-or-transformation test" is not the exclusive test, but it can still be a useful one for inventions that actually meet it. Those that do not may qualify under an alternate theory, although the courts do not provide much guidance on formulating such a theory. A general theory, if one exists at all, remains elusive, prompting one to believe that patent eligibility may be more of a case of "I'll know it when I see it" than a defined rule, at least for inventions in the fields of medical diagnostics, computer-implemented processes, and business methods. Inventors should not be inhibited however from applying for patent coverage in these fields, since numerous inventions in each of these fields have indeed been found to be patent eligible. Inventors who are creative enough to innovate will be creative enough to present their cases before a patent examiner or tribunal in such a way that shows them to be patent eligible.

---

[29] *Bilski et al. v. Kappos, supra*, slip opinion p. 16.

Chapter **10**

Selected Topics
in Patent Strategy

Developing a patent portfolio is itself a business strategy, but strategy also plays a part in deciding what to seek patent coverage for, when and how to seek it, when and against whom to assert it, and how to defend it when its validity or enforceability is challenged. Strategic decisions are made from the time of filing a patent application through the time that the application spends in the PTO prior to grant and throughout the life of a patent after grant. Inventors, employers, attorneys, and investors will all benefit from well-thought-out strategies, as will anyone seeking to derive value from or defend a patent. This chapter addresses two acts performed at a patent's earliest stages that involve strategic planning. These stages are the preparation and filing of a patent application, and the strategic acts are the use of provisional patent applications and the drafting of claims.

10.1 PROVISIONAL PATENT APPLICATIONS

The filing of a **provisional patent application** prior to a **utility patent** application is probably the strategic tactic most commonly used by inventors and their employers. A provisional patent application is a document containing a description of an invention and filed in the PTO but not processed by the PTO other than being assigned an application number and filing date. If a utility (nonprovisional) patent application on the invention described in the provisional is filed within a year of the filing of the provisional and expressly claims the benefit of the provisional, the provisional's filing date can serve as the effective filing date of the utility application and thus of any patent that is granted on the utility application. If no utility application is filed by the 1-year date, the provisional's filing date is no longer available for use, and the provisional loses its entire value. Since a provisional application itself will not result in a patent regardless of its merit, a provisional application is not truly an application for a patent but rather a kind of placeholder for a utility patent application.

First to File: Patents for Today's Scientist and Engineer, First Edition. M. Henry Heines.
© 2014 the American Institute of Chemical Engineers, Inc. Published 2014 by John Wiley & Sons, Inc.

The strategic options presented by a provisional patent application are many. Several of these options are due to one's ability to obtain a filing date from the PTO quickly and inexpensively by filing the provisional. The quickness is due to the fact that a provisional application is not required to contain all the parts of a utility patent application such as claims or an abstract or any of a utility patent application's organization and formatting requirements. The quickest provisional applications are those known as "cover-sheet provisionals," a term used to describe documents prepared for other purposes and then submitted, with essentially nothing more than a cover sheet, as a provisional application. The cover sheet is most often a transmittal form downloaded from the PTO website and filled in with such information as the identification of the inventor(s), a correspondence address, and the filing fee. The low expense of a provisional is in part because a provisional has a lower filing fee than a utility application: for applicants not qualifying for discounts (see Appendix A for the qualifications and the amounts of the discounts), the filing fee for a provisional is a single fee of $400 compared to the three-part fee (base filing fee, search fee, and examination fee) totaling $1600 for a utility application. A utility application incurs even higher fees when it exceeds certain size limits in terms of the number of pages and the number of claims (also shown in Appendix A), although these excess page and claim fees are subject to discounts as well. (Note that a utility application filed after, and claiming the benefit of, a provisional application will require the full utility application filing fees with no credit for fees paid with the filing of the provisional.) In addition to lower filing fees, provisionals often entail lower attorney fees, or even no attorney fees for those applicants who prepare provisionals on their own, particularly those filing cover-sheet provisionals. When an attorney is used, the drafting of claims and an abstract and the attorney's involvement in the organization and formatting of the text all contribute to the attorney's time, in addition to the attorney's own contribution to the strategy of the application and the patent that will issue on it.

Obtaining an early effective filing date for a utility (nonprovisional) application by first filing a quickly prepared provisional application is certainly useful in a priority dispute between utility applications under the first-to-file rule of the AIA, but it offers further benefits as well. One of these is providing an early prior art cutoff date for a later utility application, and another is serving as a prior art date itself when the utility application (or patent) that relates back to it is cited as prior art against another utility application. For purposes of cutting off prior art, that is, predating documents or activities that would otherwise qualify as prior art, a well-kept laboratory notebook can be useful, as explained in Chapter 4, but provisional applications offer benefits that laboratory notebooks either cannot provide or can only provide if combined with other facts or showings. For example, an inventor seeking to predate prior art by submitting a notebook entry as evidence of an earlier date of invention must also show that in the time between the notebook entry and the filing of the patent application the inventor has diligently continued to pursue the invention, either by working further on the invention itself or by working with a patent attorney on the patent application. In contrast, the inventor who can predate prior art with the filing date of a provisional application can do so without the need to show diligence. Nor is there any need for the inventor or the inventor's employer to preserve the record of a provisional, other than to know its application number and filing date so that these can be referenced in the nonprovisional application, since once filed the provisional is officially on record at the PTO. In litigation, for example, inventors who wish to introduce a notebook entry into evidence are typically called upon to testify as to how the notebook was preserved to protect it from alteration and often to also provide a witness to corroborate the date and authenticity of the entry. No such testimony is needed with a provisional application, since its date and authenticity are verified in the PTO record, and there is no opportunity for modifying it after it is submitted. When the situation is reversed and a utility patent or published application that claims the benefit of a provisional application assumes the status of prior art, the provisional's filing date becomes the effective prior art date. Thus, for example, a utility application filed on October 1, 2014, with no benefit from a provisional application can be rejected over a utility application filed later, for example, on November 1, 2014, that claims benefit of a provisional application filed earlier, for example, on January 1, 2014.

Another strategy associated with the filing of a provisional application is the postponement of the need to make a choice between revealing the invention to the public and maintaining the invention as a trade secret. A provisional application is initially held in confidence by the PTO but is ultimately made available to the public if a utility application with a claim to the provisional is filed within the 1-year deadline and allowed to be published, since the utility application will be published whether or not a patent is ultimately granted. If no utility application is filed (or if one is filed and then withdrawn before it is published), the official record of the provisional remains closed and the contents of the provisional remain confidential, unless the provisional or its contents have otherwise been made public, by the inventor or the inventor's employer, for example. The confidentiality of the invention in a provisional that has been allowed to lapse with no utility application having been filed is therefore preserved, but the lapsed provisional provides no protection against competitors and its filing date does not become the effective date of any patent, including those serving as prior art. Often, the need for and advantage of an early filing date may not become known until well after both the provisional application and the utility application have been filed. The potential of an advantage however usually justifies the early filing of a provisional, regardless of when the decision regarding a utility application (to file or not to file, or to pursue to grant or to abandon) is made. The effective filing date of the utility application is secured at the provisional stage.

A further strategy is the use of the provisional to extend the time between the effective filing date and the expiration date of a patent. If a utility application is filed in a timely manner and succeeds in becoming a patent, the expiration date of the patent will be 20 years from the filing date of the utility application, and in some cases longer if delays attributable to the PTO occurred during pendency, rather than 20 years from the filing date of the provisional. A utility application can thus be filed as late as the lapsing date (the 1-year anniversary) of the provisional and still claim the benefit of the provisional's filing date, and if this is done, the time span between the prior art cutoff date and the patent expiration date will be 21 years. Without the provisional, the time span is only 20 years. Note however that the provisional does not extend the term of enforceability of the patent, since a patent is not enforceable until it issues; instead, the provisional shifts the term into the future by the length of time between the filing dates of the provisional and the utility application, that is, a maximum of 1 year. The shift occurs without loss of an early prior art cutoff date, and this can be an advantage for inventions whose licensing opportunities or royalty income does not typically occur in the first year of enforceability. This is true of inventions whose commercial use is preceded by years of clinical or field testing or for which market development requires an extended period of time. (Note that a patent does not have a 20-year term, as is often mistakenly said, since the term begins with the grant of the patent, which is typically well more than a year after the filing date of the utility application. Again, however, extensions can occur if certain types of delays occurring prior to grant are attributable to the PTO rather than the applicant.)

Yet another strategy associated with provisional patent applications is the option that a provisional offers the applicant of postponing the final decision on whether to fully commit to the financial investment involved in obtaining patent coverage. A provisional application filed at the conception stage of an invention affords the inventor a year to test the concept for efficacy or to determine, verify, or modify the scope of the concept. From a strictly financial standpoint, inventors filing provisional applications have a year to raise funds from potential investors before committing to the filing fees and attorney fees incurred in filing a utility application, and the same is true for inventors wishing to determine whether their inventions have sufficient industrial or commercial appeal to justify the costs. Postponing the decision for the year can be done without loss of patent rights if the prior art cutoff date is a determining factor of patentability.

With any of these strategies, care must be taken to avoid a false sense of security from the filing of a provisional application. Even if a provisional application is filed early enough to predate prior art that might be cited against a subsequent utility application, the provisional must be adequate to support the claims of the utility application and therefore effective to overcome the prior art. Certain provisionals, for example, are filed at a stage where the inventor has not yet progressed beyond a rudimentary concept of the invention, and a provisional that is hastily written and filed at this stage may contain only the bare outlines of the concept with little or no explanation of how to implement it. The provisional may lack the detail needed for fully describing how to make and use the invention or the scope needed to support the broader claims of the utility application that is subsequently prepared in less of a hurry. Cover-sheet provisionals are particularly vulnerable to these deficiencies. Those whose contents are lifted from technical papers, for example, frequently focus on experiments actually performed by the authors with little or no statement of broader fields of application. Many technical papers also include statements that compromise patent coverage, such as negative conclusions, statements of doubt or inadequacies of the work performed, and recommendations for further testing to verify results or to confirm conclusions. Provisionals that are essentially trade show presentations repackaged for filing in the PTO are often similarly deficient since their purpose is more to attract interest among investors and consumers than to present information in a way that will support a broad patent claim. They thus tend to stress the advantages of a new product or service while withholding the underlying technical details that are required in a patent application.

Even with a full description of technical details, a provisional application may fail to provide sufficient descriptive content to support the strategies that are often needed for achieving grant of a patent application, defending the validity of the patent against a post-grant challenge, or allowing the patent to realize its full potential as a competitive instrument. These strategies can include the ability to shift to alternate ways of expressing the invention to assert its novelty when facing prior art that the inventor was not initially aware of or to trim the scope of the invention when the claims are held to be overbroad. The strategy-minded inventor or patent owner will therefore try to predict the various ways in which a patent can be challenged and of the various persons or entities, including those in different segments of the industry, over whom the patent owner might eventually wish to assert the patent. Drafting the provisional with these in mind can contribute considerably to its value.

Many patent owners adopt the habit of preparing every patent application as a fully drafted utility application and filing it as a provisional application and then refiling it as a utility application at the 1-year date. As noted earlier, this results in a patent with an expiration date that is 1 year later than it would be if the application were originally filed as a utility application, and yet it eliminates any question of support in the provisional for the claims in the utility application. This procedure offers no advantage however in the quickness by which a provisional application can be prepared, nor does it offer the financing flexibility of a provisional. Nor is it strictly necessary for assuring proper support. The courts have repeatedly stated that "[i]t is not necessary that the [earlier] application describe the claim limitations exactly … but only so clearly that persons of ordinary skill in the art will recognize from the disclosure that appellants [i.e., the inventor(s)] invented processes including those limitations."[1]

The need for a provisional application to contain enough description to provide a flexible strategy in the ultimate patent was demonstrated in a case involving Patent Nos. 5,899,283 and 5,950,743 of New Railhead Manufacturing, L.L.C.[2] Both patents were applied for on the same date and claimed the benefit of the same provisional application that had been filed 9 months earlier. The claims of the '283 patent were apparatus claims and those of the '743 patent were method claims, but the subject matter of both was horizontal, or "trenchless," drilling. Trenchless drilling is used for oil and gas exploration in stratified earth formations and also for forming boreholes underneath roadways and bodies of water for the placement of utility conduits. Trenchless drilling typically requires a drill string terminating in a directable drill bit and a device known as a "sonde" that sends positioning data to a receiver on the

[1] *In re* Wertheim, 541 F.2d 257, 263 (CCPA) (1976).

[2] "Drill Bit for Horizontal Directional Drilling of Rock Formations," U.S. Patent No. 5,899,283, issued May 4, 1999, and "Method for Horizontal Directional Drilling of Rock Formations," U.S. Patent No. 5,950,743, issued September 14, 1999, both naming Cox as inventor and assigned to New Railhead Manufacturing, L.L.C.

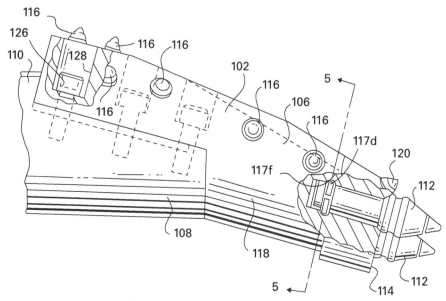

Figure 10.1   Selected figure from Patent No. 5,899,283.

ground. High-speed fluid jets are used both to steer the drill bit and to cool the drill body and blade during drilling.

Prior to the inventions in the New Railhead patents, trenchless drilling systems had been unable to drill through rock, even with the help of the fluid jets. To address this problem, the inventor in the two patents had devised an asymmetric drill bit whose body was angled relative to the sonde housing and drill string. A drawing from the '283 patent is shown in Figure 10.1.

As the drawing shows, the body of the drill bit (102) had a heel (the left half of the drill bit) extending above the axis of the sonde housing (110) and a toe (the right half of the drill bit) extending below the axis, with end studs (112) protruding from the tip of the toe. (The axis itself is not shown but is horizontal and runs along the longitudinal centerline of the cylindrical sonde.) In use, the bit rotated intermittently in a random, orbital motion that allowed the studs to fracture the rock at one location and then move to another location at a different angle relative to the axis of the sonde housing to resume fracturing. The rotation of the angled bit body relative to the axis produced a borehole whose radius exceeded that of the drill bit itself. The claims of the '283 patent were drawn to the drill bit and included the phrase "the unitary bit body being angled with respect to the sonde housing." The claims of the '743 patent were drawn to "A method of horizontal directional drilling in rock" and included the phrase "causing a drill bit at one end of a drill string to intermittently rotate as it digs in, stops rotation until the rock fractures, and then moves in a random, orbital inter-mittent motion." The method claims did not expressly recite the angled orientation of the drill bit, but New Railhead conceded that the method of the '743 claims was performed whenever the drill bit of the '283 claims was used.

When the patents issued, New Railhead sued its competitors Vermeer Manufacturing Company and Earth Tool Company for patent infringement, and the defendants responded by challenging the validity of both patents. The challenge succeeded at the trial court, which therefore declared the patents invalid, and this decision was affirmed on appeal.[3] The challenge to the method claims (the '743 patent) was based on grounds that did not call the provisional application into question, but the challenge to the drill bit claims (the '283 patent) did, asserting that the provisional's filing date could not be applied to those claims because the provisional did not show or make reference to the angled orientation of the drill bit. The provisional's filing date was essential to the drill bit claims because New Railhead had made a sale of drill bits within the scope of the claims on a date 6 months before the provisional's filing date but 15 months before the '283 patent's filing date. Thus, unless the provisional's filing date could be applied to the claims in the '283 patent, the claims would be invalidated by an on-sale bar, that is, a bar to patentability due to a sale of units within the scope of the claims more than 1 year before the effective filing date of the patent in regard to those claims (see Chapter 3). Use of the provisional's filing date would have reduced the time between the sale and the filing date to less than a year, avoiding the bar.

The application on which the '283 patent was granted, and hence the patent itself, described the angled relationship between the bit body and the sonde housing in its text and showed it in its drawings, specifically the drawing reproduced in Figure 10.1. The closest descriptions that could be found in the provisional application, on the other hand, were a "high angle of attack," "asymmetrical geometry," "offset drill points," and a "trailing shoe." The provisional did not state that the bit body was at an angle to the axis of the sonde housing, and the drawings in the provisional application showed the bit body and the sonde housing in separate views but not joined. When questioned at trial, the inventor testified that if one were to construct the parts from the drawings and combine them, the resulting drill bit would be angled as claimed in the '283 patent, but this contradicted the inventor's own deposition where he stated that there was nothing in the provisional application that indicated that the bit was angled. After considering all this, the appeals court ruled that the provisional did not adequately support the claims of the '283 patent and affirmed the trial court's decision that the patent was invalid because of the sale of units 15 months before the patent's filing date.

10.2   STRATEGIES IN CLAIM CONSTRUCTION

The exclusionary power of a patent, that is, its ability to allow the patent owner to exclude others from practicing the invention and to grant licenses to do so, resides in the claims, which thereby form the core of a patenting strategy. The claims delineate in explicit terms the metes and bounds of the products or activities that constitute

[3] *New Railhead Manufacturing, L.L.C. v. Vermeer Mfg. Co. and Earth Tool Company, L.L.C.*, 298 F.3d 1290 (Fed. Cir. 2002).

infringement of the patent. While the patent is in force, therefore, the claims are its most important part, despite their location at the end of the patent and the absence of any heading, special type fonts, or much else to set them apart from the body of the patent (the specification) or draw attention to them. In fact, the placing of the claims after the bulk of the description in the patent can often mislead the reader as to the scope of the patent's coverage, since the descriptive sections of the patent often suggest a broader scope of coverage than the claims themselves.[4] Nevertheless, the claims provide the arguments and options often needed for guiding the patent application through the examination process at the PTO, for asserting the patent against infringers as a competitive tool, and for providing the litigator with legal strategies when asserting and defending the patent in court. While this book does not attempt to instruct scientists, engineers, and other inventors how to draft their own claims, much less their own patent applications, the strongest and most valuable patents in terms of both content and strategy are those that are prepared through collaboration between the inventor and the attorney. For this reason, inventors will benefit from understanding certain basics of claim construction and in particular how to draft a claim to best serve its different functions and to provide the strategies that will give the patent its greatest value.

Most patents have figures in the form of drawings, graphs, or diagrams to supplement the specification and aid the reader's understanding of the invention. The figures generally depict specific examples or embodiments of the invention, however, while the claims express the invention as a whole, using generic terms to provide breadth of coverage. For this reason, the claims are intended to be read and understood without reference to the figures. As an exercise, therefore, read the following claim without reference to the drawings that appear on the succeeding page, and try to form a mental image in a generic sense of the physical object (a clothespin) that the claim describes. This claim is the leading claim of an actual, although expired, U.S. patent, reproduced verbatim (including punctuation and spelling errors) and in the format in which the inventor's attorney presented it to the PTO and in which it appears in the patent. (Hint: if you become lost before reaching the end of the claim, you will be in the majority.)

Claim 1: A clothespin comprising a pair of generally symmetrical, longitudinally, elongated, generally rectangular, block-like non-metallic members each having a longitudinal axis and similarly configured inner and outer surface portions; each of said pair of elongated members including a forward clamping jaw having its inner surface portion adapted for gripping a clothesline and articles to be hangably suspended therefrom when the clamping jaw of one of said pair of members closes into clamping engagement with the corresponding clamping jaw of the other of said pair of members, a tapered tail adapted to be manually engaged for levering said forward clamping jaws open, and an intermediate portion between said jaw and said tail for defining a fulcrum about which each of said pair of members may be pivoted for opening and closing said clamping jaws with respect to one another; metal spring

---

[4] For an illustration, see M. Henry Heines, *Patents for Business—The Manager's Guide to Scope, Strategy, and Due Diligence*, Westport, CT: Praeger Publishers, pp. 21–24 (2007).

means for pivotally urging said clamping jaws toward one another in a clamping position, said spring means including a coiled hollow body portion intermediate first and second relatively straight elongated end portions; the intermediate portion of each of said pair of members including a transverse lateral groove across the inner surface portion thereof, said groove being generally perpendicular to said longitudinal axis and having a generally semi-circular cross-sectional configuration such that the coiled hollow body portion of said spring means may be houseably received between the correspondingly opposed lateral grooves in the inner surface portions of said pair of elongated members and at least partially concealed therein, the intermediate portion and tail of one of said pair of members having a first relatively straight longitudinal recess in the inner surface portion thereof for receivably retaining and recessably concealing said first relatively straight elongated end portion of said spring means and the intermediate portion and tail of the other of said pair of members having a secong relatively straight longitudinal recess in the inner surface portion thereof for receivably retaining and recessably concealing said second relatively straight elongated end portion of said spring means so that the end portions of said spring means are recessed beneath said inner surface portions and said coiled body portion is at least partially recessed to prevent or at least minimize metal contact with said clothesline and said suspended articles, a first metal clip means for retainably securing one of said pair of members to the coiled hollow body portion of said spring means and a second metal clip means for retainably securing the other of said pair of members to the coiled hollow body portion of said spring means so as to prevent the inadvertent separation of said members and said spring means even under abnormal conditions, each of said members including a three-sided continuous slot formed in the external surfaces of the intermediate portion thereof and extending from one end of said lateral groove in the inner surface portion along the side of said intermediate portion, laterally across the outer surface portion of said intermediate portion directly opposite said lateral groove and along the opposite side of said intermediate surface to the opposite end of said lateral groove, the outer surface slot portion being generally parallel to said lateral groove and perpendicular to said longitudinal axis and said side slots being generally perpendicular to said lateral groove; each of said clip means being an integral piece of resilient metallic material generally configured as a partially opened rectangle when in operative position, said clip means having a relatively straight intermediate clip body, a pair of relatively straight clip sides which are generally perpendicular to said clip body and a pair of clip ends disposed toward one another and generally parallel to said clip body and perpendicular to said clip sides, said body portion being adapted to be receivably retained and recessably concealed within said outer surface slot portion, said clip sides being adapted to be receivably retained and recessably concealed within said slot sides, and said clip ends adapted to be insertably received and retained within the opposite hollow ends of said coiled spring body thereby preventing separation while simultaneously recessing said metal clip means beneath the exterior side and outer surface portions of said members to prevent or at least minimize metal contact with said clothesline and the articles suspended therefrom to prevent rust spots and the like.

Now, read the following, which is the same claim with no changes to its wording except for the removal of a misplaced comma and the correction of a spelling error, but with added paragraphing and indentation and the use of italics to highlight the physical components of the clothespin. While reading this version of the claim, refer to Figure 10.2, which contains selected drawings from the patent including the patent's own drawing representing the prior art (Part E of Fig. 10.2). Numbered parts from the drawings have been inserted in the claim to show how the claim elements are exemplified in the drawings.

Claim 1 (with paragraphing and part numbers): A clothespin comprising

a. *a pair of generally symmetrical, longitudinally elongated, generally rectangular, block-like non-metallic members* (*Part A: 12, 13*) each having a longitudinal axis (*17*) and similarly configured inner and outer surface portions (*inner: 18, 27; outer: 19, 28*); each of said pair of elongated members including

  (i) a forward *clamping jaw* (*Part B: 22, 31*) having its inner surface portion (*23, 32*) adapted for gripping a clothesline and articles to be hangably suspended therefrom when the clamping jaw of one of said pair of members closes into clamping engagement with the corresponding clamping jaw of the other of said pair of members,

  (ii) a tapered *tail* (*Part A: 24, 33*) adapted to be manually engaged for levering said forward clamping jaws open, and

  (iii) an *intermediate portion* (*Part B: 25, 34*) between said jaw and said tail for defining a fulcrum (*26*) about which each of said pair of members may be pivoted for opening and closing said clamping jaws with respect to one another;

b. metal *spring means* (*Part C: 14*) for pivotally urging said clamping jaws toward one another in a clamping position, said spring means including a coiled hollow body portion (*35*) intermediate first and second relatively straight elongated end portions (*36, 37*);

the intermediate portion (*25, 34*) of each of said pair of members including a transverse lateral *groove* (*Part A: 39, 41*) across the inner surface portion thereof, said groove being generally perpendicular to said longitudinal axis and having a generally semi-circular cross-sectional configuration such that the coiled hollow body portion (*35*) of said spring means may be houseably received between the correspondingly opposed lateral grooves in the inner surface portions of said pair of elongated members and at least partially concealed therein,

the intermediate portion and tail of one of said pair of members having a first relatively straight longitudinal *recess* (*Part A: 42*) in the inner surface portion thereof for receivably retaining and recessably concealing said first relatively straight elongated end portion (*36*) of said spring means (*14*) and the intermediate portion and tail of the other of said pair of members having a second relatively straight longitudinal recess (*43*) in the inner surface portion

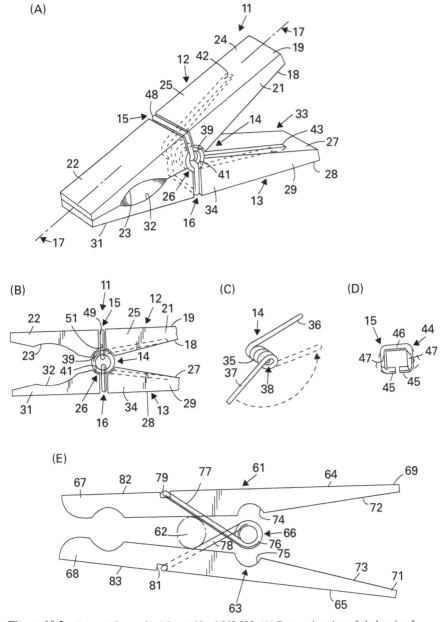

Figure 10.2 Selected figures from Patent No. 4,063,333. (A) Perspective view of clothespin of invention. (B) Fragmentary side view of (A). (C) Perspective view of spring. (D) Front view of clip. (E) Prior art as presented in the patent.

thereof for receivably retaining and recessably concealing said second relatively straight elongated end portion (*37*) of said spring means so that the end portions (*36, 37*) of said spring means are recessed beneath said inner surface portions and said coiled body portion (*35*) is at least partially recessed to prevent or at least minimize metal contact with said clothesline and said suspended articles,

c. a *first metal clip means* (*Part D: 15*) for retainably securing one of said pair of members to the coiled hollow body portion of said spring means and a *second metal clip means* (*16*) for retainably securing the other of said pair of members to the coiled hollow body portion of said spring means so as to prevent the inadvertent separation of said members (*12, 13*) and said spring means (*14*) even under abnormal conditions,

each of said members including a three-sided continuous *slot* (*Part A: 48*) formed in the external surfaces of the intermediate portion thereof and extending from one end of said lateral groove (*39, 41*) in the inner surface portion along the side of said intermediate portion, laterally across the outer surface portion of said intermediate portion directly opposite said lateral groove and along the opposite side of said intermediate surface to the opposite end of said lateral groove, the outer surface slot portion (*49*) being generally parallel to said lateral groove (*39, 41*) and perpendicular to said longitudinal axis (*17*) and said side slots (*51*) being generally perpendicular to said lateral groove; each of said clip means (*15*) being an integral piece of resilient metallic material generally configured as a partially opened rectangle when in operative position,

said *clip means* having

(i) a relatively straight intermediate *clip body* (*46*),

(ii) a pair of relatively straight *clip sides* (*47*) which are generally perpendicular to said clip body and

(iii) a pair of *clip ends* (*45*) disposed toward one another and generally parallel to said clip body and perpendicular to said clip sides,

said body portion (*46*) being adapted to be receivably retained and recessably concealed within said outer surface slot portion (*49*), said clip sides (*47*) being adapted to be receivably retained and recessably concealed within said slot sides (*Part B: 51*), and said clip ends (*45*) adapted to be insertably received and retained within the opposite hollow ends of said coiled spring body (*35*) thereby preventing separation while simultaneously recessing said metal clip means (*15, 16*) beneath the exterior side and outer surface portions of said members (*12, 13*) to prevent or at least minimize metal contact with said clothesline and the articles suspended therefrom to prevent rust spots and the like.

The second version of the claim is easier to follow than the first, and this benefits the strategic considerations listed earlier, since the inventor, the PTO, and any attorneys seeking to assert the patent against infringers or defend it against challengers will need to know from the claim wording itself exactly what structures fall within the scope of the claim and what structures do not. Also, to promote a more efficient and thorough examination of the pending application, the PTO itself encourages attorneys to incorporate paragraphing and indentation into their claims so that the claims are easier to follow.[5] Even in its revised form, however, the length of the claim raises questions, not least for the reason that the invention is a clothespin, a structurally simple device whose defining features should presumably be expressible in a claim that is correspondingly simple in both language and length.

Regardless of how simple or complex the invention, the optimal claim sets forth the metes and bounds of the invention with just enough specificity to distinguish the invention by its point(s) of novelty, rather than describing in detail the drawings that are included to illustrate the invention. Since each word or phrase in a claim is a *limitation* on the scope of the claim, the longer the claim, the more limitations it has and the narrower its scope.[6] The goal is therefore to include as few limitations as possible, with the understanding that certain limitations will be needed to put the others in a proper context. Claims of narrow scope do have a certain appeal, of course, since they will most likely encounter fewer rejections from a patent examiner and also be less likely to overlap with (and therefore be invalid over) prior art that might be discovered after the patent is granted. The extreme example of a narrow-scope claim is a "picture claim," which is a claim that lists all details of the patent's drawings. A picture claim is therefore a verbal picture of the unit that the inventor has actually constructed, listing all, or essentially all, of the features of the unit regardless of their criticality to the functionality or novelty of the concept at the core of the invention. (A method claim can also be a "picture claim" for analogous reasons, for example, by listing all materials, equipment, steps, operating conditions, amounts and proportions, etc., regardless of their criticality and their relation to the point of novelty.) As the ultimate narrow-scope claim, a "picture claim" is usually the easiest to distinguish over the prior art, since it contains so many limitations that any number of them can be pointed to as needed when prior art is cited against it. Many inventors whose patents contain only picture claims however realize too late that a competitor's product that includes what the inventor considered to be her/his point of novelty falls outside the scope of the claims by sidestepping one or more of the limitations that are peripheral to the point of novelty. A patent whose broadest claim is a "picture claim" is therefore relatively easy for a competitor to design around, and picture claims provide a very limited scope of coverage, often more limited than the invention deserves.

[5] The United States Patent and Trademark Office, Nonprovisional (Utility) Patent Application Filing Guide, January 2012: "Each claim should be a single sentence, and where a claim sets forth a number of elements or steps, each element or step of the claim should be separated by a line indentation." http://www.uspto.gov/patents/resources/types/utility.jsp#heading-17

[6] This is true for the majority of claims in utility patents, but not all. The use of the word "comprising" (the third word in the claim), commonly used in apparatus and method claims, is one reason why the breadth of the claim is inversely proportional to the number of elements and characteristics listed in the claim.

Returning to the clothespin claim, therefore, how did the claim get to be this long? And must the claim be as narrow as it appears to be (contain as many limitations as it has)? First, however, let us look at the clothespin itself as represented by the drawings in Figure 10.2 and at what the specification says about the novelty and nonobviousness of the invention.

As the drawings show, the clothespin of this invention has only five parts: two wooden ("nonmetallic") pieces (elements 12 and 13 of Part A of Fig. 10.2) that together form the clothespin's clamping jaws as well as its finger grips, a central spring (element 14 of Parts A, B, and C), and two U-shaped clips (only one of which, element 15, seen in Parts A, B, and D, is assigned a number) that secure the spring to the wooden pieces.

Turning to the specification, the problem addressed by the invention and the solution that the invention presents are stated in the opening paragraphs as follows:

> *The prior art teaches many clothespins of the type having a pair of elongated clamping members and a spring interposed therebetween for normally urging the jaws of the clothespin closed upon one another. Many of the springs used in the prior art quickly lose their resiliency and render the clothespin ineffective after a short period of use. Helical springs have proven to be extremely effective at providing a sufficient clamping force over an extended period of use.*
>
> *However, many of the clothespins of the prior art utilizing helical springs are structurally unstable. If a clothesline or an article of clothing or the like is inserted between the jaws of the clothespin to a point sufficiently close to the fulcrum, the body of the spring may lift out of the grooves provided therefore causing the clothespin to fall or fly apart if any type of lateral or longitudinal force is applied to its tail portion.*
>
> *Many of the prior art attempts to solve this problem involve the use of sliding sheaths or staple arrangements whereby the spring is physically attached at one or more points to the two halves of the clothespin. This increases the complexity and cost of the clothespin so as to render many such proposed modifications to be economically unfeasible.*
>
> *Furthermore, since most of the springs utilized in the prior art are metal, they are subject to rust. Therefore, the clothesline or any article of clothing or the like which is inserted between the jaws of the clothespin and allowed to physically contact the spring may become soiled or acquire a rust deposit often necessitating re-laundering of the article.*

The invention therefore addresses two problems—the instability of the mounting of the helical (coil) spring to the clamping members and the risk of soiling the clothing that is held between the jaws of the clothespin due to contact of the clothing with the rusting spring. The invention in claim 1 addresses the first of these by adding the two clips ("first metal clip means" and "second metal clip means," one of which is shown in Part D of Fig. 10.2) to hold the helical spring against the clamping jaws and the second by mounting the spring in the opposite direction so that the elongated ends of the spring extend toward the rear of the clothespin (the finger-grip or tail end) ("the intermediate portion and tail of one of said pair of members having a first relatively straight longitudinal recess in the inner surface portion thereof for receivably retaining and recessably concealing said first relatively straight elongated end portion of said spring means," as expressed in the claim—emphasis added) rather than toward the clamping end[7] (compare Part A of Fig. 10.2 to Part E, which represents the prior art).

Returning to claim 1, the claim as originally presented to the PTO was less than half as long as the version shown here. The official PTO record shows that the added wording, which includes various modifiers throughout the body of the claim plus the entire section at the end of the claim, beginning with "each of said members including a three-sided continuous slot …," was added by amendment in response to the first Office Action. The amendment was made in response to a rejection of the claim for obviousness over two pieces of prior art, a 1931 patent entitled "Clamp or Clothespin" (No. 1,810,304) and an 1873 patent entitled "Clothes Clamps" (No. 140,659). The clothespin depicted in the 1931 patent was identical to the clothespin that was shown in the figure labeled "prior art" (Part E of Fig. 10.2) in the application being examined.

Accompanying the amendment was the following explanation (the comments in parentheses are added herewith):

> [N]either of the references teach a coiled hollow spring body having relatively straight elongated end portions (the 1873 patent actually did show a spring meeting this description); semi-circular lateral grooves for houseably receiving the coiled spring body (the 1931 patent actually did show this) and relatively straight longitudinal recesses along the inner surface of the intermediate and tail portions for retainably receiving and recessably concealing the straight end portions of the spring to prevent clothes-metal contact (these features were already in the claim prior to the amendment) … [nor did either reference] show or suggest Applicant's exterior recess along the sides and outer surface of the intermediate portion opposite the lateral groove for receivably retaining and recessably concealing the clip members which secure the elongated members to the spring assembly while recessing the metal clips below the surface of the member to prevent clothes-metal contact (in fact, neither reference showed any "clip members" at all). None show Applicant's claimed clip configuration where the elongated body portion is disposed in an external lateral recess and the opposing clip ends enter the hollow coil body portion.

Was it necessary to more than double the length of the claim in response to the rejection, and is the resulting claim a "picture claim"? Specifically, are all the limitations in the claim necessary to the novelty and nonobviousness of the clothespin (compare Part A of Fig. 10.2 to Part E, the acknowledged prior art) or necessary to the functionality of the elements of the clothespin (i.e., are they needed to place the elements in their proper context so that the claimed invention is still a clothespin)? The following is a list of limitations from the claim selected with these questions in mind.

---

[7] Other claims in the same patent recite additional features (limitations) of the clothespin that further protect the clamped clothing article from contact with the rusting spring. The claim addressed here however is an independent claim that does not recite these additional limitations but does state that the limitations that it does recite prevent at least some contact with the spring.

Claim element	Modifier	Remarks
The clamping members	"generally symmetrical" "generally rectangular" "block-like"	Do any of these modifiers distinguish over the prior art? Are the symmetry and the profile critical or necessary to the functionality of the clamping members?
The "tails" (finger-grip ends of the clamping members)	"tapered"	Does the tapered shape distinguish over the prior art? Must the tail of the clamping member be tapered for the user to be able to open and close the jaws?
"Recess" in the tail of each clamping member to receive the elongated ends of the coil spring	"in the inner surface portion" "recessably concealing said ... elongated end portions [of the coil spring]" "recessed beneath said inner surface portions ... to prevent or at least minimize metal contact with said clothesline and said suspended articles"	Must the ends of the spring contact the inner surfaces of the tails? Metal contact with the suspended articles is avoided by moving the ends from the front (clamping) half of the clamping member to the rear half (the tails), which can be achieved regardless of whether the retaining recesses are on the inner or outer surfaces of the tails. Also, is it critical to make the recesses deep enough to conceal the ends of the coil spring, considering that both the clothesline and the articles will be between the clamping ends (jaws) of the clamping members rather than between the tails?
The clips ("clip means")	"of resilient metallic material" "configured as a partially opened rectangle ... having ... straight intermediate clip body, a pair of ... straight clip sides" "clip ends ... parallel to said clip body and perpendicular to said clip sides"	Prior art: must the clips contain all of these features to distinguish the invention over the two references, neither of which includes clips? Functionality: must the clips be resilient and metallic to hold the spring in place? The "resilient" feature is most likely for ease of assembly, but assembly will be done by the manufacturer rather than the user. The "metallic" feature is most likely for strength, but other materials would function as well. Must the actual physical shape of each clip be specified since the claim also states in the last paragraph that the clips are "insertably received and retained within the opposite hollow ends of said coiled spring body" (other shapes may achieve the same result)?

The "slot" for each clip	"three-sided" (referring to the rectangular slot profile as indicated in Part B of Fig. 10.2) "extending from one end of said lateral groove [i.e. half of the cylindrical recess surrounding the coil section of the spring] ... along the side of said intermediate portion [of the clamping member], laterally across the outer surface portion of said intermediate portion directly opposite said lateral groove and along the opposite side of said intermediate surface to the opposite end of said lateral groove"	Prior art: are these features needed to distinguish the invention over the two references, neither of which includes clips, much less slots for clips? Functionality: is the profile of each slot critical to the slot's function of holding the clip in place? Can it be semicircular and achieve the same result? Must the groove extend all the way around the clamping member to stabilize the clip? Is a groove necessary at all, considering that the clothing article suspended by the clothespin will be held between the clamping ends and will not contact the clips?

Removing one or more limitations in the "Modifier" column will broaden the claim (the more limitations removed, the greater the broadening), and admittedly this will render the claim more vulnerable to challenge for lack of novelty or for obviousness over prior art, particularly prior art that may be discovered after either the application filing date or after the patent issue date. This vulnerability can be addressed however by placing the removed limitations, individually or in combinations, in dependent claims to serve as backup positions. Doing so will provide strategic options that are not available with the claim as shown.[8] In the broadest claim, the modifiers should be those that are needed for context, clarification, or distinction over the prior art or all three. Those that limit a component to aspects of appearance or shape that are neither critical to the component's functionality nor relevant to its distinction over the prior art should be avoided. With these considerations in mind, an approach to drafting the broadest claim is to first identify the problem that the invention is intended to solve; then isolate and find words to express the features that embody the inventor's solution to the problem, that is, the point(s) of novelty (which

---

[8] The patent does have three dependent claims, claims 2, 3, and 4, referring back to claim 1 (the independent claim quoted earlier), but each adds only the feature of a "pin means" (i.e., a pin) mounted on the clamping jaw half of one of the clamping members and extending toward and into a hole in the opposing member, the pin serving as an alternate means of avoiding contact between the clothing article and the spring by traversing the gap between the two clamping members, thereby limiting how far the clothing article can be inserted into the gap. A second independent claim, claim 5, focuses on the pin itself as the point of novelty, without limiting the orientation of the spring or including any mention of the clips.

is also nonobvious), without limiting the features more than is needed to express their functionality; and finally add only those additional features that are needed to place the point(s) of novelty in a context that will reflect their functionality. The resulting claim will express the invention in an economy of words and yet have the breadth of scope that the invention justifies.

How many points of novelty (and nonobviousness) should the broadest claim recite? If a single limitation is both novel and nonobvious, this is sufficient to make a claim allowable and enforceable. Multiple points can certainly be included, but novelty is not a matter of degree, and hence a claim with multiple points of novelty does not meet the novelty requirement any better than a claim with only one. Nonobviousness may be considered a matter of degree, but insurance against insufficient nonobviousness in the leading claim can be obtained by the use of dependent claims. The clothespin claim shown here, which is the first claim in the patent, recites both the "clip means" and the rearward orientation of the elongated ends of the coil spring (extending toward the finger-grip ends of the clamping members)—two points of novelty in the same independent claim. The specification, as quoted earlier, indicates that each of these features addresses a distinct problem and is able to do so independently of the other. There may be a certain degree of interaction between these two features—for example, the need for the clips as stabilizers for the spring may be greatest when the spring ends are in the rearward orientation. Nevertheless, each feature has at least some effect in lessening the problem it is directed to without the inclusion of the other feature. This suggests that further flexibility in terms of asserting the patent over competing clothespin manufacturers could be achieved by placing these two features in separate independent claims. This would further increase the overall scope of coverage by encompassing manufacturers that incorporate only one of the two features as well as those that incorporate both. A third independent claim that recites both features could still be retained as backup support.

Chapter 11

Patents and Beyond: The Variety and Scope of Intellectual Property

Utility patents, which are the main focus of this book, can certainly be an important part of any company's assets and a valuable means of establishing a company's position in a competitive industry. Other assets however of a similarly intangible nature can add value as well. Trade secrets, trademarks, copyrights, and patents other than utility patents, when appropriate to a company's business model or product line, are major contributors to the valuation and competitiveness of many types of business. Patents, trade secrets, trademarks, and copyrights are all forms of "intellectual property" (IP), which can be defined as any information, skill, or ability, technology-based or not, that is the product of creative imagination and is in a form that can be recognized as proprietary and used as a means of conferring exclusivity. IP is thus distinguishable from the conventional business structures, models, and platforms that are commonly used in industry, such as management hierarchies, hiring practices, marketing methods and consumer targeting, or the manner in which research and development priorities are set. An understanding of the basics of forms of IP other than utility patents will help place utility patents in their proper context.

11.1 TRADE SECRETS

Other than patents, trade secrets are the form of IP that is best suited for protecting technology. Unlike patents, however, one does not apply for trade secret status to an agency of either a state government or the federal government; there is no state or federal "Trade Secret Office." Companies generate trade secrets as part of their daily business routine, either by labeling documents containing technical information as confidential or by instituting a company trade secret policy and encouraging

First to File: Patents for Today's Scientist and Engineer, First Edition. M. Henry Heines.
© 2014 the American Institute of Chemical Engineers, Inc. Published 2014 by John Wiley & Sons, Inc.

employee loyalty. Trade secrets do not generally require special procedures for their creation, but they can be strengthened by informing and reminding all company personnel of the importance to the company of internally developed information and of the obligation of each employee to preserve that information. An express obligation is typically included in the employment agreement that all new hires are asked to review and sign on their first day on the job. Memoranda or other internal announcements or communications that formally notify all employees that company information is confidential and must be protected are also a means of establishing and reinforcing a trade secret policy.

Trade secrets can cover the same types of technological innovations that are eligible for patent protection, but they can also go much farther since they are not limited, as are patentable inventions, to information that is "novel" and "nonobvious." Any information that is both private and useful, that is of value to the company, and that if released to competitors or to the public at large may lower the value of the company's product line or its competitive advantage can qualify as a trade secret. Trade secrets can thus extend to such information as choices of materials, sources of supply, customer lists, and specialized marketing plans and business strategies.

While it is most often a company's employees who are bound by its trade secrets, those who are not employees can be bound as well. Business partners such as outside equipment manufacturers, suppliers, consultants, independent contractors, governmental agencies, and even customers are frequently expected and expressly obligated to honor trade secrets that may be revealed to them through their routine transactions with the trade secret owner. Trade secret obligations are thus often found in purchase orders, contracts, and other types of documentation that set forth the mutual obligations and interests of the parties.

The expiration date of a trade secret can be set by agreement, but trade secret status can continue indefinitely if proper procedures are followed to ensure its preservation. Aside from any agreed-upon expiration date, a trade secret is extinguished (i.e., its trade secret status is lost) by public disclosure, whether the disclosure occurs in violation of an obligation to maintain the trade secret or by someone who is under no such obligation. Extinguishment can occur, for example, when a newly hired employee discloses to the new employer a trade secret of the former employer, either inadvertently or intentionally, intentional disclosure occurring perhaps due to resentment to having been terminated or to having been denied a promotion. Extinguishment can also occur by an inadvertent or careless disclosure at an industry convention or exhibition. Trade secret extinguishment by public disclosure that is neither careless, malicious, or otherwise in violation of an obligation can occur, for example, by an individual who has arrived at the same information independently by the individual's own creative efforts and then published the information or by an individual who uncovers the information by reverse engineering a product released by or legitimately obtained from the owner, particularly when there was no obligation on the individual's part not to reverse engineer.

Violation of a trade secret occurs when the trade secret has been "misappropriated," and misappropriation can occur upon the unauthorized *acquisition, use,* or *disclosure* of the secret. Those who *acquire* the secret knowing or having reason to know that they did so improperly have thus violated the trade secret and can be held liable to the trade secret owner. *Disclosure* or *use* of a trade secret without the express or implied consent of the owner can result in liability if those disclosing or using the secret either did so in violation of an obligation not to or acquired it improperly, knowingly obtained it from someone who acquired it improperly, knowingly obtained it from someone who was obligated to maintain its secrecy, or acquired it by accident or mistake and became aware that it was a trade secret before they took advantage of it, such as by making an investment in a way that would benefit from the trade secret or entering into a contract involving use of the trade secret. Certain ways of acquiring trade secret information without the owner's consent are permissible, however, and do not constitute misappropriation. These include independently inventing or discovering the information, with or without publishing it but without knowing that someone else considered it a trade secret, reverse engineering the trade secret as mentioned earlier, with or without disseminating the knowledge, observing the trade secret in public use or on display, and obtaining the information through published literature that is accessible to the industry or to the public at large. Once acquired in one or more of these ways, the information can be freely disclosed or used by those acquiring it without liability to the trade secret owner.

When properly preserved, trade secrets can be of considerable value to a company. Even preserved, however, trade secrets have certain limitations. One is the scope of their subject matter, and the other is the public policy of not recognizing trade secrets that effectively prevent individuals from practicing their professions and from using their technical skills to do so.

11.1.1 Scope

The scope of a trade secret is typically determined by the form in which the trade secret information arises or, when the information is documented, the contents of document. Unlike patents, trade secrets are not generally created with the intention of staking out territory that the company wishes to claim as proprietary, as in patents, but instead with a particular focus, either as part of a specific project or marketing plan, for example, or for solving a particular problem or to be part of, or used in or for, a particular product or service. How far the secret extends beyond that focus or the context in which it was generated is often a matter of argument depending on whose interest is at stake, that is, the owner or the one accused of misappropriation. In general, the further one deviates from the information itself or its original focus, the more difficult it is to claim trade secret misappropriation.

11.1.2 The Right of an Individual to Use Fundamental Skills

An employee's obligation to preserve a present or past employer's trade secrets is a legitimate interest and right of the employer. All industries recognize however that individuals should have the freedom to change employers and still use their skills and experience. Skills acquired prior to the start of employment, such as those learned at a university or a vocational school, are typically expanded and improved on the job as an employee gains experience and learns ways to increase productivity, efficiency, and product quality. This type of skill development makes the employee more valuable to the employer but is also part of the employee's basic learning curve. Differentiating between acquired knowledge that is more properly characterized as an addition to or expansion of one's professional skill set and that which is company proprietary is at times difficult, however. When an employer sues a former employee for trade secret misappropriation, therefore, the tribunal will often seek to balance the interest in encouraging the use of trade secrets to promote investment in research and sustainable business activity with the need to allow the individual to work in her/his chosen field. Trade secret obligations that effectively prevent the individual from doing any work in the same field can thus be limited by a court of law and in extreme cases declared entirely unenforceable.

11.1.3   Comparing Trade Secrets to Patents

Trade secrets differ from patents in many respects, prominent among which are scope of coverage, duration of enforceability, the disclosure of proprietary information to the public, and cost. For both trade secrets and patents, the scope of coverage can be a matter of choice, but in most cases, only patents are specifically prepared with the intention of defining and obtaining coverage for a particular scope. As for duration, the term of a patent is limited by statute, while a trade secret can remain in force indefinitely. A patent is by definition a public disclosure of the technical information that it protects, while a trade secret once publicly disclosed is no longer a trade secret. In terms of cost, patent coverage has significant upfront costs, including both PTO filing fees and attorney fees, as well as additional PTO fees and attorney fees between filing and issuance, depending on what the application encounters during examination. Still further costs, in the case of utility patents, are the maintenance fees that the PTO imposes to keep the patents in force once they are issued. Trade secrets, by contrast, do not entail official fees of any kind and can be established without the services of an attorney. The public disclosure distinction between patents and trade secrets is of particular concern to certain types of business for which the potential for harm to a company's interests from public disclosure of the information through a patent is considered to be outweighed by the power conferred by the patent to threaten competitors with infringement liability. Since a trade secret owner can restrict and thereby control the dissemination of its trade secret information while the information in a patent is publicly available and readily searchable, even before the patent is granted, companies with this view tend to prefer trade secrets. Both patents and trade secrets however have their place among the corporate IP.

It is thus common business practice among technology companies to maintain both a patent portfolio and trade secrets. Rolls-Royce Corporation is one such company, and its dispute with AvidAir Helicopter Supply, Inc., illustrates Rolls-Royce's ability to use trade secrets to protect proprietary information beyond that which is covered by its patents. The dispute was related to helicopter engines, and

Rolls-Royce has a sizable portfolio of U.S. patents on inventions of use in helicopter engines, including patents that were acquired from the Allison Engine Company when Allison was purchased by Rolls-Royce. The particular engines that were the subject of the dispute were the Model 250 engines, a commercially successful family of turboshaft engines with a reverse airflow configuration. The engines were first developed and manufactured by Allison in the 1960s, and production was continued by Rolls-Royce when the acquisition occurred in 1995.

According to procedures that were established soon after the Model 250 engines were first introduced, Allison's customers when seeking repair or overhaul of the engines took the engines to any of various independent overhaul shops. These shops serviced the engines using procedures that were developed by Allison and certified by the Federal Aviation Administration (FAA). Updates of the procedures were periodically made by Allison and each update was certified by the FAA. At first, the overhaul shops were freely able to obtain the procedures and updates from Allison, but Allison eventually instituted a policy of designating certain overhaul shops as authorized maintenance centers (AMCs) and providing the updates only to the AMCs. In exchange for receiving an update, each AMC was required to sign a confidentiality agreement specifying that the information in the update was proprietary to Allison. The agreement prohibited the AMC from disseminating the information to persons or companies outside the AMC and required the AMC to return all proprietary documents to Allison if and when the AMC was no longer authorized. Rolls-Royce's acquisition of Allison occurred the year after the new policy took effect, at which time the obligations of the AMCs became due to Rolls-Royce, which continued the policy.

AvidAir Helicopter Supply, Inc., was not an AMC but nevertheless entered the repair and overhaul market the same year that the new policy took effect, using procedures that were then current. Several years and several updates later, Rolls-Royce noticed that AvidAir was not following the procedures in the most recent update and filed a complaint with the FAA. Upon inspection of AvidAir's procedures, the FAA agreed with Rolls-Royce and ruled that AvidAir was not using a certified procedure. AvidAir then obtained a copy of the most recent update from a source other than Rolls-Royce, modified its procedure to conform to the update, and certified to the FAA that it was now in compliance. Still lacking AMC status, however, AvidAir obtained the updates and made the modifications without Rolls-Royce's permission. Rolls-Royce then sent AvidAir a cease-and-desist letter, and AvidAir responded by bringing a legal action against Rolls-Royce to have the technical information in the update declared unenforceable as a trade secret and to have Rolls-Royce's denial of permission declared a violation of antitrust laws and an interference with AvidAir's business interests. Rolls-Royce then filed its own lawsuit against AvidAir for trade secret misappropriation. The validity of the trade secret was the principal issue in both cases, which were consolidated into a single case in the Western District of Missouri. The decision at the trial court was in Rolls-Royce's favor and was affirmed on appeal to the U.S. Court of Appeals for the Eighth Circuit.[1]

---

[1] *AvidAir Helicopter Supply, Inc. v. Rolls-Royce Corporation*, 663 F.3d 966, (8th Cir. 2012).

The two lawsuits were originally filed in Missouri and Indiana, respectively, under the Uniform Trade Secrets Act (UTSA), which had been adopted by these 2 states and 45 others plus the District of Columbia. According to the UTSA, a trade secret is defined as:

> *information, including a formula, pattern, compilation, program, device, method, technique, or process, that: (1) derives independent economic value, actual or potential, from not being generally known to, and not being readily ascertainable by proper means by, other persons who can obtain economic benefit from its disclosure or use; and (2) is the subject of efforts that are reasonable under the circumstances to maintain its secrecy.*

It was acknowledged that parts of the overhaul procedure included in the update that AvidAir was using were also included in earlier updates that were not protected by the AMC agreements. Other parts of the update were new, however. The overall procedure was thus a compilation of old and new parts, and as the UTSA definition indicates, compilations are one of the types of information that the UTSA recognizes as qualifying for trade secret status. Noting also that the update afforded the AMCs a competitive advantage and that the information it contained was not readily ascertainable by other means, the court found that the update met part (1) of the UTSA's definition of a trade secret. As for part (2), AvidAir argued that the ease with which it was able to obtain the information indicated that Rolls-Royce had not taken reasonable steps to keep the information secret. The court disagreed, noting that all of the documents included in the update were labeled with proprietary-rights legends and that none of the documents had been distributed to anyone who was not bound by a confidentiality agreement. The court also noted that AvidAir had acquired the documents from others who were either not authorized to provide them or who had themselves misappropriated them. The fact that AvidAir was able to obtain them so easily did not defeat the fact that Rolls-Royce had made reasonable efforts to maintain their secrecy. Nor was the court persuaded by the relatively wide distribution of the documents: there were 28 AMCs and the documents had been distributed to each one.

AvidAir also argued that the current update should not be afforded trade secret status since it contained only a small amount of information that was not also available from prior (nonconfidential) updates and that the new information offered no technical advances over the old. The court responded by drawing attention to the fundamental difference between patents and trade secrets, which is that patent law awards property rights based on novelty and nonobviousness, while trade secret law governs commercial ethics by recognizing rights that are based on economic value and secrecy. A trade secret thus derives its value from being a secret rather than from the merit of any technical improvement it may contain. As a result, Rolls-Royce was awarded $350,000 in damages and a permanent injunction requiring AvidAir to discontinue repairing and overhauling Model 250 engines and to return the protected documents to Rolls-Royce.

11.2 TRADEMARKS

A **trademark** is a commercial source identifier and takes the form of a word, symbol, or device that is identified with the goods of a particular manufacturer or distributor and that distinguishes those goods from the goods of other manufacturers or distributors. **Service marks** and **trade names** are also commercial source identifiers and are used in ways analogous to trademarks. Like a trademark, a service mark associates a service with the service provider. Similarly, a trade name is the name of a corporation or other business entity that symbolizes the identity and reputation of that entity aside from the merits of specific goods or services that the company provides. Trademarks, service marks, and trade names are thus valuable forms of IP, conferring property rights and protection for a range of subject matter that differs considerably from that of patents. For convenience, the term trademark is used here in a collective sense to include all three—trademarks, service marks, and trade names.

11.2.1 Choosing a Trademark

Terms and symbols for use as trademarks are generally chosen at the early stages of developing a corporate identity or a marketing strategy for a product or service. The best marks are those that have market appeal but are also both available for use and protectable. To be available for use, the mark must be one that is not similar to a mark already in use by another, and to be protectable, a mark must be of the type that is enforceable against infringers.

Establishing trademark availability typically begins with a search for similar marks, followed by an analysis to determine if the proposed mark might raise a "likelihood of confusion" (the legal standard for trademark infringement) with any of the marks produced by the search. Thus, if the public is likely to confuse suppliers of different goods or services because one supplier's mark is confusingly similar to that of the other, the supplier who adopted the mark earlier will be entitled to enforce the mark against infringers and the other will not.

To be protectable, a mark must meet several requirements. One is that it must not be merely descriptive of the product, since this would prevent others from using common words to describe their products. Another is that the mark be one that is not deceptively misdescriptive, since this would mislead the public as to the nature of the product. A third requirement is that the mark not be merely a surname or a geographical term, unless it can be shown that the mark has acquired a secondary meaning, that is, a distinctiveness that is independent of the person whose surname it is or the geographical region that bears the same term, in the minds of the public. The most protectable marks are therefore those that are coined terms that have no inherent linguistic meaning. Words that do have a common meaning can also be protectable if they are arbitrary, that is, if they have no connotation in association with the goods they are used on. In some cases, words that are suggestive of particular qualities of the product or service can be valid marks, provided that they do more than merely describe the product or service. In general, the less arbitrary the words or the less inherent linguistic meaning the words have, the greater their need for secondary meaning if the words are to be protectable.

11.2.2   Securement, Maintenance, and Infringement of Trademark Rights

Trademark rights are created simply by use of the mark in commerce. Neither state nor federal registration of a mark is required for the mark to be an enforceable trademark, although both are available. Federal registration in particular offers certain benefits, including a geographical scope beyond the actual area over which the user's reputation has extended and the ability to claim the right to a trademark based on a "bona fide intent to use" the mark rather than actual use. The term of enforcement of a trademark can be unlimited, although to maintain a federal trademark registration, a declaration of continued use (or excusable nonuse) must be filed with the USPTO at 5- to 10-year intervals.

Proper handling of a trademark will ensure its viability as an enforceable property right. Proper handling includes (i) avoiding use of the mark as a generic term and instead using it in a manner that clearly indicates that it is a brand; (ii) using special typography for the mark, such as all capital letters or a distinctive typeface, to alert the public to its use as something other than a generic or descriptive word; (iii) using a trademark notice, such as "TM" or "SM" for trademarks and service marks, respectively, and a circled "R" for federally registered marks, to alert the public of the proprietary nature of the mark; and (iv) policing the use of the mark by others to prevent uncontrolled and unauthorized uses.

Trademark infringement occurs when a product of a supplier other than the trademark owner is labeled with a mark that is close enough to the trademark that it raises a *likelihood of confusion* in the mind of the consumer as to whether the source of the product is the product's supplier or the trademark owner. Such confusion would benefit the infringer by giving the product the undeserved advantage of the trademark owner's reputation. Confusing similarity is thus the most common ground for liability to a trademark owner. Trademark owners will also seek to prevent their marks from being "diluted" by the registration of similar marks by others and from becoming "generic" by being used by others as a descriptive term rather than a trademark. Those who refer to the marks of others are therefore advised to do so in such a way that the trademark status of the marks and the identity of their owners are clear.

11.3   COPYRIGHTS

Copyrights protect original works of authorship and are most commonly applied to music, art, books, and other published literature. Business-oriented materials are also protectable by copyright, however. Computer software, instruction manuals, technical data sheets, and logos are examples.

A copyright protects the expression of an idea rather than the idea itself. Marketing materials, for example, may be protectable by copyright while the underlying marketing concept is not; a written description of a bookkeeping system may be protectable, while the bookkeeping system itself is not. The right conferred by a copyright is the right to prevent others from copying, distributing, and publicly performing or displaying the copyrighted work and from creating derivative works, that is, works based on the copyrighted material. A derivative work in a business context may thus be a translation of the underlying work into a different language, an adaptation of the underlying work for a different medium, an editorial revision of the work, or an annotated form of the work.

A copyright is created by fixing the work in a tangible medium of expression that is readable either directly or with the aid of an electronic or other device. Copyright protection can thus be obtained not only by publishing the work in print media or recording media but also by placing it on the Internet.

The duration of a copyright, for works created on or after January 1, 1978, and that are not "made for hire," extends from the date of creation of the copyright to 70 years after the death of the author. Copyrights on works made for hire have durations that last until 95 years from the year of first publication or 120 years from the year of creation, whichever is earlier. Works "made for hire" are those prepared by an employee within the scope of employment or those specially ordered or commissioned for various purposes set by the Copyright Act, including translations, instructional texts, contributions to collective works, and the like. Unlike other copyrighted works, a work made for hire is owned by the employer or other party who ordered or commissioned the work.

Copyrighted works typically contain a copyright notice that includes the symbol © plus the year of first publication and the name of the copyright owner. While once a requirement for asserting a copyright over an infringer, this notice is no longer required for works published on or after March 1, 1989. The notice is still recommended, however, since it puts infringers on notice that they will be subject to copyright liability if they copy the work.

Copyrights are most often registered with the U.S. Copyright Office, which is part of the Library of Congress, although like the copyright notice, registration is not a requirement for copyright protection. Registration, particularly when done promptly after creation of the work, offers important advantages, however. These include the right to bring a lawsuit against an infringer, a presumption of validity of the copyright in a court of law, and the ability to collect damages for copyright infringement that are set by statute.

Copyright infringement occurs when any of the various acts listed earlier, that is, copying, distributing, etc., are done without the copyright owner's permission. "Fair use," such as criticism, comment, news reporting, teaching, scholarship, and research, can be a defense to a claim of copyright infringement, but easy access is not. Placing a copyrighted work on the Internet, for example, does not deprive the work of copyright protection.

11.4 DESIGN PATENTS

Like copyrights, **design patents** protect esthetic features rather than utilitarian features. Design patents are nevertheless patents and are only obtained through application to the USPTO and only on designs that meet the requirements set forth in the patent statute.[2] By virtue of these requirements and published decisions rendered in courts of law in design patent cases, design patents differ considerably from copyrights in terms of how they are infringed, how long they are enforceable, and the requirements for their validity.

While copyrights are directed to works of authorship, design patents are directed to articles of manufacture and protect the "visual ornamental characteristics embodied in, or applied to, an article of manufacture ... the configuration or shape of an article, [or] the surface ornamentation applied to an article."[3] The description of a patentable design, according to the statute, is a "new, original, and ornamental design," as compared with the statute's description of a patentable invention as a "new and useful process, machine, manufacture, or composition of matter." A patentable design must also be non-obvious, as must a patentable invention, but nonobviousness is much less well defined in design patents and highly subjective. The standard is the "ordinary observer," a parallel to the "person having ordinary skill in the art" of utility patents, but the use of the term "ordinary observer" emphasizes the visual nature of the design rather than any utilitarian function of the article that the design is applied to or associated with.

A design patent has only one claim, and it refers to the drawings that are the critical part of the patent, the patent having essentially no specification other than a list of the drawings. The scope of coverage of a design patent is thus determined by its drawings, and infringement occurs in the same way as it does for a utility patent, except with reference to the drawings rather than to the claims. While copying is a requirement for infringing a copyright, it is not a requirement for infringing a design patent. Thus, making, using, selling, etc., an article that embodies or contains the same design shown in the patent's drawings, whether copied from the drawings or not, will infringe. Design patents do have scope, however, and certain designs that are not identical to those shown in the drawings will also infringe. The standard for determining whether the patent is infringed by a nonidentical design is whether the "ordinary observer with knowledge of the prior art designs" would find the accused design to be "deceptively similar" to the design shown in the patent drawings.[4] Thus, the "ordinary observer" serves a role in determining both nonobviousness (and therefore patentability) and infringement. The current definition in the law of an "ordinary observer" is less amorphous than that of the "person having ordinary skill in the art" in utility patents: the law defines an "ordinary observer" as one who is familiar with all of the relevant prior art. One wonders however whether a hypothetical person can meet this definition and still be "ordinary."

---

[2] Title 35 of the United States Code, Sections 171–173.

[3] The United States Patent and Trademark Office, Design Patent Application Guide. Available at http://www.uspto.gov/patents/resources/types/designapp.jsp. Accessed August 13, 2012.

[4] *Egyptian Goddess, Inc., et al., v. Swisa, Inc., et al.*, 343 F.3d 665 (Fed. Cir. 2008).

The term of a design patent is 14 years from its date of issuance, regardless of the application's filing date, and the fees associated with design patents are considerably lower than those for utility patents, for both filing and issuance. The attorney fees will also be much lower, considering that the essence of the patent is its drawings. As noted earlier, however, the scope and focus of the typical design patent are considerably narrower than those of the typical utility patent. Nevertheless, many technology companies apply for design patents to supplement their utility patent portfolio or to obtain a form of patent protection for a new and useful product that is not sufficiently nonobvious to qualify for a utility patent.

11.5   IP COVERAGE FOR PLANTS

A highly specialized field of IP law of interest to plant breeders, growers, and gardeners is that of **plant patents** and **plant variety protection certificates**. Like utility and design patents, plant patents are obtained by application to the USPTO, and they cover plant varieties that are either developed or discovered by the applicant and that have been asexually reproduced, with the explicit exceptions of tuber-propagated plants and plants found in an uncultivated state.[5] Thus, sports, mutants, hybrids, and transformed plants can be covered by plant patents, the sports or mutants being either spontaneous or induced and the hybrids being either natural, obtained through a planned breeding program, or somatic in source. Bacteria cannot be covered by plant patents, but algae and macro fungi can.

Like a design patent, a plant patent contains only one claim, and it refers to the drawings and the specification, which is required to contain a botanical description of the plant. A plant patent is infringed by one who asexually reproduces the plant, uses the plant, offers it for sale, or sells it or any of its parts. Like utility patents, the term of a plant patent is 20 years from the filing date of the application on which the patent was granted.

For sexually reproduced plants, tuber-propagated plants, and F1 hybrids, protection can be obtained in the form of a plant variety protection certificate, obtained through application to the **Plant Variety Protection Office**, an agency of the Department of Agriculture. Similar to a patent, a plant variety protection certificate gives its owner the right to exclude others from selling the variety, offering it for sale, reproducing it, importing or exporting it, or using it in producing a hybrid or different variety from it. The term of a plant variety protection certificate is 20 years from the date of issue of the certificate or, in the case of a tree or vine, 25 years from the date of issue.

11.6   CONCLUSION

Like patentable inventions, designs, and plants, information that is capable of protection as trade secrets or trademarks or by copyright is often not recognized as such. Unlike patents, however, the legal protections afforded by trade secrets, trademarks,

---

[5] Title 35 of the United States Code, Sections 161–164.

and copyrights often come into existence simply upon the creation of the information, marks, or works themselves. Recognizing the value of any form of IP is a major first step toward preserving it and its contribution to a company's assets.

A comparison chart summarizing the differences between the various forms of IP discussed in this chapter is presented in Tables 11.1, 11.2, and 11.3.

Table 11.1   Intellectual property comparison chart, Part 1: trade secrets and trademarks

	Trade secrets	Trademarks
Qualifying subject matter	Any information that is of value to its owner and that if released may lower the value of its owner's product or reduce its owner's competitiveness	A word, symbol, or device that is identified with the goods of a manufacturer or distributor and distinguishes those goods from the goods of competitors
How to obtain	By agreement or company policy	By use, although strengthened by state or federal registration (through the U.S. Patent and Trademark Office)
Duration	Until expired according to the terms stated in the agreement; otherwise unlimited until the information is publicly disclosed	Unlimited, although requires periodic declarations of continued use
How to infringe	By misappropriation, defined as use, acquisition, or disclosure with knowledge of its status as a trade secret	By use of a mark so similar to the trademark that the public may be confused into thinking that the trademark holder is the source of the goods bearing the similar mark

Table 11.2   Intellectual property comparison chart, Part 2: copyrights and utility patents

	Copyrights	Utility patents
Qualifying subject matter	Any original work of authorship but only the work's particular expression of the idea expressed in the work rather than the idea itself	Any "process, machine, manufacture or composition of matter" (per the U.S. Patent Statute) that is novel, useful, and nonobvious
How to obtain	By fixing the work in a tangible medium of expression, including the Internet	For U.S. patents, by application to the U.S. Patent and Trademark Office; for patents outside U.S., by application to individual countries or to multicountry authorities established by treaty

Table 11.2   (Continued)

	Copyrights	Utility patents
Duration	Until 70 years after the death of the author if not made for hire; if made for hire, either 95 years from the first publication of the work or 120 years from the creation of the work, whichever comes first	From grant date to expiration date, which is 20 years from the effective filing date, with extensions under limited circumstances; requires periodic maintenance fees after grant to remain in force throughout term
How to infringe	By unauthorized copying, distribution, or public performance or display and by creating derivative works	By making, using, selling, offering for sale, and importing the patented subject matter

Table 11.3   Intellectual property comparison chart, Part 3: design patents, plant patents, and plant variety protection certificates

	Design patents	Plant patents	Plant variety protection certificates
Qualifying subject matter	Visual ornamental characteristics embodied in, or applied to, an article of manufacture	Asexually reproduced plant varieties, except for tuber-propagated plants and plants found in an uncultivated state	Sexually produced plants, tuber-propagated plants, and F1 hybrids
How to obtain	By application to the U.S. Patent and Trademark Office	By application to the U.S. Patent and Trademark Office	By application to the Plant Variety Protection Office
Duration	14 years from date of issuance	From grant date to expiration date, which is 20 years from the filing date	From grant date to expiration date, which is 20 years from the issue date
How to infringe	By making, using, selling, offering for sale, or importing the patented subject matter	By asexually reproducing, using, selling, or offering for sale the plant or any of its parts	By selling, offering for sale, reproducing, importing, exporting, or using the variety

Appendix A

Selected Fees Charged by U.S. Patent and Trademark Office and Other U.S. Agencies for Intellectual Property as of January 1, 2014

The tables below include fees charged by government agencies for patents, copyrights, trademarks, and plant variety protection certificates. These tables are not complete lists of the fees charged by these agencies, but rather lists of fees selected as those most likely to be of immediate interest to individuals and their employers or to those providing financial support to individuals in securing these various forms of intellectual property protection. For a complete list of patent and trademark fees, the reader is advised to consult the U.S. Patent and Trademark Office (USPTO) website at http://www.uspto.gov/web/offices/ac/qs/ope/fee010114.htm, for copyright fees, the Copyright Office website at http://www.copyright.gov/docs/fees.html, and for plant variety protection certificate fees, the USDA website at http://www.ams.usda.gov/AMSv1.0/ams.fetchTemplateData.do?template=TemplateR&page=PlantVarietyProtectionOfficeItemizedFees&description=Itemized%20Fee%20List. Better yet, consult a patent attorney or other appropriate intellectual property practitioner, since fees for certain submissions to these agencies and at certain stages throughout their procedures have multiple components that are not readily apparent, even from a comprehensive list. One should also be cautioned that these fees change frequently, which is another reason for consultation with a practitioner familiar with the IP area of interest. Fees associated with plant variety protection certificates are payable to the

First to File: Patents for Today's Scientist and Engineer, First Edition. M. Henry Heines.
© 2014 the American Institute of Chemical Engineers, Inc. Published 2014 by John Wiley & Sons, Inc.

Plant Variety Protection Office (a division of the U.S. Department of Agriculture), and fees associated with copyrights are payable to the Copyright Office.

As the first table indicates, most fees for patents are three tiered, including a regular level and two levels of discounts. The "**small entity**" discount is available to an independent inventor, a nonprofit organization, or a business concern whose number of employees, including affiliates, does not exceed 500 and that has not assigned, granted, conveyed, or licensed (and is under no obligation to do so) any rights in the invention to any individual or entity that does not itself qualify for "small entity" status. A "nonprofit organization" is defined as a university or other institution of higher learning (in any country), a 501(c)(3) organization (per the Internal Revenue Code of 1986), a nonprofit scientific or educational organization as qualified under a nonprofit organization statute of a state of the United States, or a nonprofit scientific or educational organization of a country outside the United States that would qualify as a nonprofit U.S. organization if it were located in the United States.

The "micro entity" discount, where offered, is available to an applicant (which includes individuals and their assignees) that meets the following criteria: (i) the applicant must qualify as a "small entity"; (ii) none of the inventors can have been named as an inventor on more than four previously filed patent applications, other than provisional applications and applications filed in another country (including international applications filed under the PCT); (iii) the applicant must not, in the calendar year preceding the year in which the fee is being paid, have had a gross income (as defined by the IRS) exceeding three times the median household income for that calendar year (as reported by the Bureau of Census); and (iv) the applicant must not have assigned, granted, or conveyed, and is under no obligation to do so, a license or other ownership interest to an entity with a gross income that meets the criteria of (iii). Third-party submitters or challengers do not qualify for "micro entity" status.

"Small entity" and "micro entity" discounts are not available for trademarks, copyrights, or plant variety protection certificates.

Table A1.1 Fees charged by U.S. Patent and Trademark Office in connection with patents

Description	Fee (in U.S. dollars)		
	Regular	Small entity	Micro entity
Provisional patent application			
Filing fee for up to 100 sheets	260	130	65
Surcharge for each additional 50 sheets in excess of 100 sheets	400	200	100
Surcharge for late filing fee of provisional application cover sheet	60	30	15
Utility patent application			
Basic filing fee for electronic filing with up to 3 independent claims and 20 claims total and a total of 100 sheets (including drawings)	280	70	70
Basic filing fee for nonelectronic (paper) filing with up to 3 independent claims and 20 claims total and a total of 100 sheets (including drawings)	680	340	270
Surcharge for each independent claim in excess of three	420	210	105
Surcharge for each claim (independent or dependent) in excess of 20; multiple dependent claims are counted as multiple claims (equal to the number of claims they depend from)	80	40	20
Inclusion of multiple dependent claims (single charge)	780	390	195
Surcharge for each additional 50 sheets in excess of 100 sheets	400	200	100
Search fee	600	300	150
Examination fee	720	360	180
Surcharge for late filing fee, search fee, examination fee, or oath/declaration or filed without at least one claim or by reference	140	70	35
Surcharge for filing in language other than English	140	70	35
Request for prioritized examination	4,000	2,000	1,000
Submission of Information Disclosure Statement (prior art list), submitted by the applicant after the later of (i) the first Office Action or (ii) three months after filing the application	180	90	45
Extension of time for responding to Office Actions (communications from the examiner)—1 month	200	100	50
Extension of time—2 months	600	300	150
Extension of time—3 months	1,400	700	350
Extension of time—4 months	2,200	1,100	550

(*Continued*)

Table A1.1   *Continued*

Description	Fee (in U.S. dollars)		
	Regular	Small entity	Micro entity
Extension of time—5 months	3,000	1,500	750
Issue fee	960	480	240
Utility patent after issuance			
Maintenance of patent at 3.5 years from issuance	1,600	800	400
Maintenance of patent at 7.5 years from issuance	3,600	1,800	900
Maintenance of patent at 11.5 years from issuance	7,400	3,700	1,850
Surcharge for late payment of maintenance fee within 6 months of due date	160	80	40
Third-party challenges to pending utility patent application			
Submission of documents or information for more than three items	180	90	90
Petition for derivation proceeding	400	400	400
Third-party challenges to issued utility patent			
Post-Grant Review: at request, for up to 20 claims; for each additional claim	12,000; 250	12,000; 250	12,000; 250
Post-Grant Review: at grant of request, for up to 15 claims; for each additional claim	18,000; 550	18,000; 550	18,000; 550
Covered Business Method Review: at request, for up to 20 claims; for each additional claim	12,000; 250	12,000; 250	12,000; 250
Covered Business Method Review: at grant of request, for up to 15 claims; for each additional claim	18,000; 550	18,000; 550	18,000; 550
Inter Partes Review: at request, for up to 20 claims; for each additional claim	9,000; 200	9,000; 200	9,000; 200
Inter Partes Review: at grant of request, for up to 15 claims; for each additional claim	14,000; 400	14,000; 400	14,000; 400
Design patent application and issued design patent			
Filing fee for up to 100 sheets	180	90	45
Surcharge for each additional 50 sheets in excess of 100 sheets	400	200	100
Examination fee	460	230	115
Issue fee	560	280	140
Maintenance fees	None	None	None
Plant patent application and issued plant patent			
Filing fee for up to 100 sheets	180	90	45
Surcharge for each additional 50 sheets in excess of 100 sheets	400	200	100
Examination fee	580	290	145
Issue fee	760	380	190
Maintenance fees	None	None	None

Table A1.2   Fees charged by U.S. Patent and Trademark Office in connection with trademarks

Description	Fee (in U.S. dollars)
Application for registration by paper filing, per international class	375
Application for registration by electronic filing, per international class	325
Application for registration by electronic filing meeting specified requirements, per international class	275
Filing a statement of use, per class	100
Filing a declaration of continued use (at 5–6 years), per class	100
Application for renewal (at 10 years), per class	400

Table A1.3   Fees charged by U.S. Copyright Office for copyright registrations

Description	Fee (in U.S. dollars)
Registration by paper filing	65
Registration by electronic filing	35

Table A1.4   Fees charged by Plant Variety Protection Office in connection with plant variety protection certificates

Description	Fee (in U.S. dollars)
Filing and examination	4382
Issuance of certificate	768

Appendix **B**

Patent Searchers

The term "patent searchers" is used herein to include patent professionals who are engaged to find any published materials, including patents as well as other materials that relate to an invention that one seeks to patent or that has already been patented. Patent searchers are many and varied, ranging from individuals specializing in certain areas of technology to firms that profess a capability to search all areas of technology and boast a roster of individual searchers, each with expertise in a specific area. Many searchers, particularly those operating individually, are former patent examiners and many are present and former patent attorneys or agents, and as a result, no single individual is conversant or capable in all areas of technology. In addition to differences in technological expertise, there are variations in the scope of materials that a searcher will encompass in the search, with some searchers claiming a global scope and some claiming a scope that includes all published literature, whereas patent attorneys and other clients who use searchers repeatedly tend to know which ones are most effective for which technologies and for which types of published materials or geographical regions. Variations also lie in the manner in which the searcher presents the search results, since any search will entail some degree of analysis on the part of the searcher in selecting the items to be included in the search report and some searchers provide a more detailed analysis of the items in their reports than others. Variations also lie in the number of items listed in the search report, since some searchers, in the interest of providing the client with a manageable list of search results, will limit the number of items to those most relevant and eliminate those that are cumulative. Choosing a searcher therefore requires knowing something about the searcher and the searcher's service and knowing one's own level of expertise in patent matters, particularly in terms of how much and what type of analysis one wishes the searcher to perform. And of course, one's budgetary constraints will be a factor.

One of the most critical distinctions among searches however is the purpose of the search. Purposes vary: searches of prior art to determine the patentability of one's

First to File: Patents for Today's Scientist and Engineer, First Edition. M. Henry Heines.
© 2014 the American Institute of Chemical Engineers, Inc. Published 2014 by John Wiley & Sons, Inc.

invention and how best to claim it, searches of prior art to challenge the validity of another's patent, searches for product clearance or freedom to operate, that is, to determine whether a proposed product or process will encounter possible liability for patent infringement—all are conducted differently. For a useful search, the purpose of the search must be specified before the search is performed. The searcher should also be informed of any particular areas of concern to the person requesting the search, such as where the requester believes that the most relevant items are most likely to be found (patents vs. other published material, patents issued in the United States vs. those issued in Japan, Europe, or other parts of the world), whether the search should focus on a particular individual or company as being likely to have patents or literature in the relevant area, and what features of the requester's product or process or the requester's invention are considered points of novelty or are most likely to be relied on as distinctions over the prior art.

The searchers in the following list have been selected from information supplied by attorneys who have used them. The inclusion of any particular searcher in this list is not intended as an endorsement of that searcher. All quotes are taken directly from the websites of the individual searchers.

> PriorArtSearch.com: "[A] full-service patent search firm geared toward law firms, corporations, and investment firms." Areas of technology are listed as electrical/electronic, mechanical, medical devices, software, and business methods.
>
> www.us-patent-search.us (Intelligent Patent Services): Former inventors and officers of Stanford University Office of Technology Licensing, using expertise in "artificial intelligence" to perform patentability searching.
>
> www.cardinal-ip.com (Cardinal Intellectual Property, Inc.): Listing all technical areas and all types of patent searches, including English-based patent and nonpatent literature, as well as English abstracts of many foreign language patents and published applications.
>
> www.landon-ip.com (Landon IP): "[C]omplex patent searches for law firms, companies, universities, and inventors in the electrical, chemical, mechanical, biotechnology, and medical device fields. Our searchers hold advanced technical degrees. Many of them are registered to practice before the U.S. Patent and Trademark Office (PTO), and some are former patent examiners."
>
> www.patenthawk.com (Patent Hawk): "[A] full-service patent technical consultancy. Our mainstay service is building an invalidity position for litigation firms representing defendants via prior art search ... We also help small companies fend off patent infringement assertions."
>
> www.searchpriorart.com (Gongwell Services, Inc.): "Our core competency is prior art search specialized in Japanese patent and non-patent literature, as well as in Chinese and Korean. We use original-language databases (CJK, Chinese, Japanese and Korean)."
>
> www.mogambosolutions.com (Mogambo Solutions, LLC): Former patent examiners and scientists and engineers, listing electrical engineering,

biomedical engineering, mechanical engineering, pharmaceutical and life sciences, and business methods as areas of technical expertise.

www.nerac.com (Nerac): "[A] custom research and advisory firm for companies developing innovative products and technologies" listing among its "intellectual property services" "patentability research, prior art research due diligence support, monitoring for infringement activities, [and] monitoring of competitive patent activities."

www.evalueserve.com (Evalueserve): "[A] team of more than 2,600 professionals worldwide" specializing in life sciences and health care.

nulltiple.com (Nulltiple Discovery L.L.C.): Dr. Pradeep Gutta, an individual specializing in chemistry, biotechnology, chemical engineering, pharmaceuticals, medical devices, and food engineering.

wolffinfo.com (Wolff Information Consulting, LLC): Thomas E. Wolff, an individual specializing in chemistry and chemical engineering, specifically listing polymers and plastics; energy, gasification, and synthetic fuels; refining and petrochemistry; consumer products and health care; competitive intelligence; and web and computer applications.

www.dolcera.com (Dolcera Infomaxx Pvt., Ltd.): Using searchers in North America and Asia, listing expertise in software, semiconductor, networking, mechanical engineering, chemical engineering, pharmaceuticals, and civil engineering, with emphasis on telecommunications, enterprise software, advertising, and medical devices.

Acronym Glossary

AIA	Smith–Leahy America Invents Act of 2011		**NPE**	nonpracticing entity
			NSF	National Science Foundation
ANDA	Abbreviated New Drug Application		**PAE**	patent assertion entity
			PCR	polymerase chain reaction
CBM	Covered Business Methods		**PCT**	Patent Cooperation Treaty
C-I-P	continuation-in-part (patent application)		**PHOSITA**	person having ordinary skill in the art
EPA	Environmental Protection Agency		**PTO**	Patent and Trademark Office
			TRIPS	Trade-Related Aspects of Intellectual Property
EPO	European Patent Office			
FDA	U.S. Food and Drug Administration		**TSM**	teaching–suggestion–motivation (test for obviousness)
FIPSE	Fund for the Improvement of Postsecondary Education			
			USPTO	U.S. Patent and Trademark Office
FOIA	Freedom of Information Act			
GATT	General Agreement on Tariffs and Trade		**WIPO**	World Intellectual Property Organization
NIH	National Institutes of Health		**WTO**	World Trade Organization

First to File: Patents for Today's Scientist and Engineer, First Edition. M. Henry Heines.
© 2014 the American Institute of Chemical Engineers, Inc. Published 2014 by John Wiley & Sons, Inc.

Glossary

All elements rule. A rule for anticipation, stating that for a claim of a utility patent to be anticipated by the prior art, all of the elements in the claim must be disclosed or contained in a single item of prior art.

Anticipation. The absence of novelty of an invention over the prior art and therefore grounds for rejecting a patent application or declaring a claim of a patent invalid. See "Novelty."

Benefit of a filing date. The attribute of a later-filed patent application enabling it to use the filing date of an earlier-filed patent application, rather than the actual filing date of the later-filed patent application, as a cutoff date for determining what qualifies as prior art against the later-filed patent application. The benefit of an earlier filing date thus enables a patent application (or patent) to eliminate certain materials from consideration as prior art.

Conception. The formation of a complete idea of an invention in the mind of the inventor.

Continuation application. A refiling of a patent application with a specification identical to that of, and claiming the benefit of the filing date of, the patent application before refiling.

Continuation-in-part application. A refiling of a patent application in expanded form, either in the scope of its claims or in the content of the description in its specification, yet containing at least one claim that has full descriptive support in the specification as it existed prior to refiling. A continuation-in-part (C-I-P) application thus has the benefit of the filing date of the application prior to expansion for at least one claim, while one or more other claims may not have that benefit and instead can only rely on the actual filing date of the C-I-P as a cutoff date for prior art.

Copyright. A legally recognized right to prevent others from copying, distributing, publicly performing, or displaying an original work of authorship that is fixed in a tangible medium of expression and from creating derivative works based on the work. Although best known for protecting music, art, and books and other published literature, copyrights can also protect software and business-oriented materials such as instruction manuals, technical data sheets, and logos.

Critical date. The date 1 year before the effective filing date of a patent application. If the application is entitled to the benefit of the filing date of an earlier-filed patent application, the critical date is 1 year before the earlier filing date. The significance of the critical date arises in different contexts but most often in the context of a sale, offer for sale, publication, or commercial use of an invention by or on behalf of the inventor before the effective filing date. If the sale, offer for sale, etc., occurs after the critical date, it can often be eliminated as potential prior art; if it occurs before the critical date, it cannot be eliminated under any circumstances, and the claims

First to File: Patents for Today's Scientist and Engineer, First Edition. M. Henry Heines.
© 2014 the American Institute of Chemical Engineers, Inc. Published 2014 by John Wiley & Sons, Inc.

of the application or patent that cover the subject matter will be either rejected or declared invalid.

Derivation proceeding. A proceeding conducted within the USPTO for determining whether the invention in a patent application naming an individual as inventor was in fact derived from another individual who was not named, the proceeding being initiated by the unnamed individual. If the unnamed individual succeeds, the naming of the inventor(s) on the patent is corrected accordingly.

Design patent. A legal instrument establishing ownership rights to visual ornamental characteristics embodied in, or applied to, an article of manufacture and affording its owner the right to exclude others from making, using, selling, and other commercial exploitation of an article bearing those characteristics.

Divisional application. A refiling of a patent application with a specification identical to that of, and claiming the benefit of the filing date of, the patent application before refiling the claims of the refiled application being limited to those excluded from the original application by order of a patent examiner due to the examiner's finding that the original application had claimed multiple inventions that required different fields of search.

Double patenting. A basis for rejection of a patent application by the USPTO, due to an examiner's finding that the invention claimed in the application is also claimed in another application by the same inventor.

Double patenting, obviousness type. A basis for rejection of a patent application by the USPTO, due to an examiner's finding that the invention claimed in the application is an obvious variant of an invention claimed in another application by the same inventor.

Experimental use exception. An exception to the on-sale bar (which includes various types of commercial activity in addition to sales) to patentability of an invention, whereby if the activity otherwise causing the bar is deemed to be primarily experimental in character, the bar is not in effect and the patentability is not affected by the activity.

Filing date, actual. The date on which a patent application is submitted to the USPTO, as distinct from the date of submission of any earlier-filed patent application of which the later application claims a filing date benefit. An actual filing date can be the date of actual receipt of the application by the USPTO, or it can be the mailing date if mailing procedures specified by the USPTO are followed or the date of electronic submission if electronic filing procedures specified by the USPTO are followed.

Filing date, effective. A filing date that a patent application uses for various purposes, most notably as a prior art cutoff date that may be either the actual filing date of the application or the filing date of an earlier-filed application, if the later-filed application is entitled to the benefit of that earlier filing date.

First-to-file rule. A rule introduced to U.S. patent law by the AIA, specifying that between competing applicants separately applying for patents on the same invention, the patent will be awarded to the applicant that has the earlier actual or effective filing date.

First-to-invent rule. The prevailing rule under U.S. patent law prior to the AIA regarding competing applicants separately applying for patents on the same invention, whereby the patent would be awarded to the applicant that was the first-to-invent the invention.

General Agreement on Tariffs and Trade. A multinational agreement for harmonizing patent laws.

Incorporated by reference. In patents and prior art, the explicit citation by one document of another document as a source for a more detailed description of a feature or element mentioned in the first document.

Interference. An administrative procedure conducted before the USPTO for patents and patent applications governed by the first-to-invent rule to determine which of competing applicants separately applying for patents on the same invention was the first-to-invent and thereby is to be awarded the patent.

Joint inventors. Two or more persons who collaborate on an invention by each contributing to the conception of the invention as expressed in at least one claim of a patent or patent application on the invention.

Machine-or-transformation test. A test for eligibility of the subject matter of an invention, as opposed to the invention itself, for patenting: if a process involves the use of a machine or transforms something, the process is eligible for patenting, that is, eligible to be examined for novelty and nonobviousness. According to the Supreme Court in Bilski v. Kappos, however, the machine-or-transformation test is not the only test for patent eligibility, although inventions that meet this test need not satisfy other tests.

Micro entity. A status introduced by the AIA for a patent applicant that qualifies the applicant for a 75% discount in many fees payable to the USPTO in the course of applying for a patent. Among the qualifications are limitations on the number of patent applications that the inventor has previously filed and on the inventor's gross income in the year that the fee is paid.

Nonobviousness. One of the core requirements for patentability of an invention over the prior art. Nonobviousness relates to the quality of the difference between the invention and the prior art, and the manner in which nonobviousness is met differs among different types of inventions.

Nonpracticing entity. A business entity that obtains ownership of patents but does not engage in manufacturing or any business activity relating to the subject matters of the patents other than licensing the patents to licensees or bringing infringement litigation against accused infringers. An alternative term is "patent assertion entity."

Novelty. One of the core requirements for patentability of an invention over the prior art. Novelty is generally met by the inclusion of any difference between the invention as claimed and any single item of prior art.

Obviousness. The absence of nonobviousness of an invention over the prior art and therefore grounds for rejecting a patent application or declaring a claim of a patent invalid. See "Nonobviousness."

Office Action. An official communication to a patent applicant or applicant's representative from an examiner at the USPTO as part of the examination procedure of a pending patent application.

On-sale bar. A bar to patentability of an invention due to the placement of an embodiment of the invention on sale more than 1 year before the filing date of a patent application on the invention, which may be an effective filing date earlier than the actual filing date if the patent application qualifies for such an earlier date. The on-sale bar includes offers for sale. Whether the bar depends on the location of the sale or offer (i.e., within or outside the United States) and whether it depends on the accessibility of the knowledge of the sale to the public (i.e., whether it is secret or nonsecret) vary with whether the patent application falls under the AIA or under the preexisting law. The sale or offer that creates an on-sale bar does not refer to the sale of offer of a prospective patent or patent application, but rather to that of a product, process, or other goods or service covered by a claim in a patent or application.

Patent. A property right granted by a country that confers upon its owner the right to exclude others from practicing (making, using, selling, offering to sell, importing, and/or other acts depending on the type of patent) an invention for a limited period of time within the country that granted the right.

Patent assertion entity. An alternative term for a nonpracticing entity.

Patent Cooperation Treaty. A multinational agreement promoting worldwide harmonization of patent laws and offering patent applicants seeking to apply for patents in a multitude of countries the ability to establish a common filing date in each country by filing the application in a single receiving office.

Person having ordinary skill in the art. The hypothetical person whose judgment determines whether an invention is either obvious or nonobvious over the prior art. The qualifications or credentials of such a hypothetic person are not set by statute, and case law is not consistent as to what the proper credentials should be.

Petitioner. One who files a petition with the USPTO for action by the USPTO outside the normal patent examination procedure. The rules relating to patents under Title 37 of the Code of Federal Regulations provide for petitions by various parties, including patent applicants and challengers, for a wide variety of reasons. Of primary interest in this book are petitions to initiate a Post-Grant Review or an Inter Partes Review.

Plant patent. A legal instrument establishing ownership rights to plant varieties that have been asexually reproduced, but not including tuber-propagated plants or plants found in an uncultivated state. A plant patent affords its owner the right to exclude others from asexually reproducing the plant, using it, offering it or any of its parts for sale, or selling any of its parts.

Plant variety protection certificate. A legal instrument establishing ownership rights in sexually reproduced plants, tuber-propagated plants, and F1 hybrids. A plant variety protection certificate affords its owner the right to exclude others from selling the variety, offering it for sale, reproducing it, importing or exporting it, or using it in producing a hybrid or a different variety.

Plant Variety Protection Office. An agency of the U.S. Department of Agriculture that reviews applications for plant variety protection certificates and issues such certificates on those applications that meet its requirements.

Pre-Grant Publication. The publication of a patent application by the USPTO at approximately 18 months from the effective filing date of the application. The pre-grant publication is done independently of the examination process and is automatic, unless a request for nonpublication is made by the applicant. Publication of the application at other times can be made by the UPTO at the request of the applicant.

Prior art. Knowledge and activity in a wide variety of forms defined by the patent statute against which a claim in a patent application can be compared to determine whether the invention recited in that claim meets the requirements of novelty and nonobviousness for patentability.

Provisional patent application. A document containing a description of an invention and filed in, but not examined by, the USPTO and, provided that certain requirements are met, serving to establish a filing date for later use as the effective filing date of a later-filed nonprovisional patent application, that is, one that will be examined for possible issuance as a patent.

Publication bar. A bar to patentability of an invention due to the publication of an embodiment of the invention more than 1 year before the filing date of a patent application on the invention, which may be an effective filing date rather than the actual filing date if the patent application qualifies for an effective date. "Publication" according to this bar can occur in various forms, and the degree to which the publication is accessible to the public is a factor in determining whether the bar applies.

Reduction to practice. An implementation of an invention, typically the first such implementation, such as the construction of a prototype or a test performance of a process, often conducted to confirm that the invention functions as conceived or that it can be constructed or will perform as conceived. Such a reduction to practice is often termed an "actual" reduction to practice, as opposed to a "constructive" reduction to practice, which is simply achieved by the filing of a patent application on the invention in the USPTO in the event that no actual reduction to practice has occurred before the filing.

Reissue application. The resubmission of a patent to the USPTO by the patent owner after the patent has been granted, for purposes of changing some aspect of the patent, on the ground that through error on the part of the applicant or the patent owner the patent claims either more or less than the applicant is entitled to claim.

Respondent. One against whom a petition is filed and who has an opportunity to respond to allegations made by the petitioner in the petition or otherwise object to grant of the petition.

Right of priority. The right of a patent application to avail itself of the benefit of the filing date of an earlier-filed patent application as a cutoff date for determining what qualifies as prior art. The term is most often used when the earlier-filed patent application is one filed through the Patent Cooperation Treaty or in a country or patent examining authority outside the United States.

Service mark. A symbol such as a name, logo, or slogan that associates a service with the supplier of the service and thereby implies that the service carries the good will that the supplier has developed in its dealings with its customers.

Small entity. A status established by the USPTO for a patent applicant that qualifies the applicant for a 50% discount in many fees payable to the USPTO in the course of applying for a patent. Business concerns with 500 employees or less (including affiliates), nonprofit organizations, and independent inventors can qualify as small entities.

Smith-Leahy America Invents Act of 2011. An act passed by Congress that introduces the most extensive modifications in U.S. patent law since 1952, including the institution of the first-to-file rule.

Specification of patent. The text of a patent (or patent application) preceding the claims, in which the invention is described in a manner that enables the reader to make and use the invention throughout the scope of the invention.

Teaching, suggestion, motivation test. A test for obviousness of an invention over prior art: an invention that is arrived at by modifying the teachings of (i.e., the description in) a prior art reference is obvious over that reference if the reference also teaches (discloses) or suggests the modification or provides motivation to the person having ordinary skill in the art reading the reference to make the modification. According to the Supreme Court in KSR International v. Teleflex, however, this is not the only test for the obviousness of a modification. Nevertheless, if it is met, other tests need not be applied.

Terminal disclaimer. The disclaiming of a patent (an agreement by the patentee not to enforce the patent against infringers) beyond a date that is prior to the expiration date of the patent as set by the patent statute, thereby shortening the term of enforceability of the patent by eliminating a terminal portion of the term. A terminal disclaimer is typically filed to overcome a rejection for double patenting.

Trade name. The name of a corporation or other business concern that symbolizes the identity and reputation of that concern aside from the merits of its goods and services.

Trade-Related Aspects of Intellectual Property, including Trade in Counterfeit Goods. A multinational agreement setting forth principles, rules, and disciplines relating to intellectual property and its impact on international trade, establishing intellectual property rights, means for enforcement of those rights, and means for settling multilateral disputes over those rights.

Trade secret. Useful information relevant to its owner's business and considered by its owner to be proprietary and confidential, and that if released to its owner's competitors or to the public at large may lower the value of its owner's product line or reduce its owner's competitive advantage.

Trademark. A symbol such as a name, logo, or slogan that associates a product with the supplier of the product and thereby implies that the product carries the good will that the supplier has developed in its other dealings with its customers.

Transitional Program for Covered Business Method Patents. A program introduced by the AIA for challenging patents on business method inventions, applicable only to inventions for "performing data processing or other operations used in the practice, administration, or management of a financial product or service," the program remaining in effect only until September 15, 2020.

Troll. A pejorative term for a nonpracticing entity or patent assertion entity.

Utility (or usefulness). One of the core requirements for patentability of an invention. Utility is generally a rudimentary requirement and refers to the ability of the invention to perform the function for which it was conceived and which is stated in the patent, although the requisite ability is not necessarily to do so to an exemplary degree or even efficiently.

Utility patent. The most common form of patent and commonly referred to simply as a "patent," protecting utilitarian features of an article, composition, system, method, or process, as distinct, for example, from ornamental or other esthetic qualities of an invention.

World Intellectual Property Organization. A multinational organization established to implement the provisions of the Patent Cooperation Treaty.

Bibliography, Websites, and Blogs

BOOKS

ANSON, WESTON, and DONNA SUCHY, eds., *Fundamentals of Intellectual Property Valuation—A Primer for Identifying and Determining Value*, Chicago: American Bar Association, Section of Intellectual Property Law (2005).

ASPELUND, DONALD J., and STEPHEN LUNDWALL, *Employee Noncompetition Law*, Eagan, MN: Clark Boardman Callaghan (2013).

BARRY, CHRIS, RONEN ARAD, LANDAN ANSELL, and EVAN CLARK, "2013 Patent Litigation Study—Big cases made headlines, while patent cases proliferate," New York: PricewaterhouseCoopers, LLP (2013). Available at http://www.pwc.com/en_us/us/forensic-services/publications/assets/2013-patent-litigation-study.pdf. Accessed May 26, 2014.

BRUNSVOLD, BRIAN G., D. PATRICK O'REILLEY, and D. BRIAN KACEDON, *Drafting Patent License Agreements*, 7th ed., Bethesda, MD: Bloomberg BNA (2012).

DAVIS, AMY E., PAULA M. BAGGER, JOANNA H. KIM, and JEFFREY K. RIFFER, *Guide to Protecting and Litigating Trade Secrets*, Chicago: American Bar Association, Section of Litigation (2013).

DICKSON, MARK, ed., *Marketing Your Invention*, 3rd ed., Chicago: American Bar Association, Section of Intellectual Property Law (2009).

FINKELSTEIN, WILLIAM A., and JAMES R. SIMS III, *Intellectual Property Handbook: A Practical Guide for Franchise, Business and IP Counsel*, Chicago: American Bar Association, Section of Intellectual Property Law (2005).

GLAZIER, STEPHEN C., *Patent Strategies for Business*, 3d ed., Washington, DC: LBI Law & Business Institute (2003).

GOLDSTEIN, PAUL, *Copyright's Highway: From Gutenberg to the Celestial Jukebox*, Revised ed., Palo Alto, CA: Stanford Law & Politics (2003).

GRAHAM, CHRIS SCOTT, *Protecting Trade Secrets—Before, During, and After Litigation*, Chicago: American Bar Association, Section of Litigation (2012).

HANELLIN, ELIZABETH, ed., *Patents Throughout the World*, St. Paul, MN: West Group (2000).

HEINES, M. HENRY, *Patent Empowerment for Small Corporations*, Westport, CT: Quorum Books (2001).

HEINES, M. HENRY, *Patents for Business*, Westport, CT: Praeger Publishers (2007).

HENN, HARRY G., *Henn on Copyright Law—A Practitioner's Guide*, 3d ed., New York: Practicing Law Institute (1991).

HITCHCOCK, DAVID, *Patent Searching Made Easy: How to do Patent Searches on the Internet and in the Library*, 6th ed., Berkeley, CA: Nolo Press (2013).

KELLER, BRUCE P., and STACY A. SNOWMAN, *Conducting Intellectual Property Audits*, New York: Practicing Law Institute (1996).

First to File: Patents for Today's Scientist and Engineer, First Edition. M. Henry Heines.
© 2014 the American Institute of Chemical Engineers, Inc. Published 2014 by John Wiley & Sons, Inc.

LECHTER, MICHAEL A., ed., *Successful Patents and Patenting for Engineers and Scientists*, New York: The Institute of Electrical and Electronics Engineers, Inc. (1995).

MALSBERGER, BRIAN M., *Employee Duty of Loyalty: State-by-State Survey*, 4th ed, Arlington, VA: BNA Books (2009).

MALSBERGER, BRIAN M., ROBERT A. BLACKSTONE, and ARNOLD H. PEDOWITZ, *Trade Secrets—A State-by-State Survey*, 3rd ed., Washington, DC: BNA Books (2006).

NOTESTINE, KERRY E., KAREN E. FORD, and RICHARD N. HILL, *Fundamentals of Employment Law*, 2nd ed., Chicago: American Bar Association, Tort and Insurance Practice Section (2000).

PIENKOS, JOHN T., *The Patent Guidebook*, Chicago: American Bar Association, Section of Intellectual Property Law (2005).

RIVETTE, KEVIN G., and DAVID KLINE, *Rembrandts in the Attic*, Cambridge, MA: Harvard Business Press (2000).

ROSENBERG, MORGAN D., and RICHARD J. APLEY, *Business Method and Software Patents—A Practical Guide*, New York: Oxford University Press (2012).

SIMENSKY, MELVIN, and LANNING G. BRYER, eds., *The New Role of Intellectual Property in Commercial Transactions*, New York: John Wiley & Sons (1994).

SIMENSKY, MELVIN, LANNING G. BRYER, and NEIL J. WILCOF, eds., *The New Role of Intellectual Property in Commercial Transactions. 1998 Cumulative Supplement*, New York: John Wiley & Sons (1998).

SMITH, GORDON V., and RUSSELL L. PARR, *Valuation of Intellectual Property and Tangible Assets*, 3rd ed., New York: John Wiley & Sons (2002).

STOBBS, GREGORY A., *Business Method Patents*, New York: Aspen Publishers (2002).

STOBBS, GREGORY A., *Software Patents*, 3rd ed., New York: Wolters Kluwer Law & Business (2012).

GOVERNMENT PUBLICATIONS

"Changes to Implement the Inventor's Oath or Declaration Provisions of the Leahy-Smith America Invents Act," *Federal Register*, Vol. **77**, No. 157, Rules and Regulations, pp. 48776–48826, United States Government Printing Office (August 14, 2012).

"Changes to Implement Derivation Proceedings," *Federal Register*, Vol. **77**, No. 176, Rules and Regulations, pp. 56068–56092, United States Government Printing Office (September 11, 2012).

Design Patent Application Guide, The United States Patent and Trademark Office, August 13, 2012. Available at http://www.uspto.gov/patents/resources/types/designapp.jsp. Accessed May 27, 2014.

"Examination Guidelines for Implementing the First Inventor to File Provisions of the Leahy-Smith America Invents Act," *Federal Register*, Vol. **78**, No. 31, Rules and Regulations, pp. 11059–11088, United States Government Printing Office (February 14, 2013).

Leahy-Smith America Invents Act, Public Law 112–29, enacted September 16, 2011, United States Government Printing Office. Available at http://www.gpo.gov/fdsys/pkg/BILLS-112hr1249enr/pdf/BILLS-112hr1249enr.pdf. Accessed May 27, 2014.

Manual of Patent Examining Procedure, 8th ed., Revision 9, Irvine, CA: Matthew Bender (2012). Available at http://www.uspto.gov/web/offices/pac/mpep/index.html. Accessed May 27, 2014.

WEBSITES

American Bar Association Section of Intellectual Property: www.americanbar.org/aba.html

The Intellectual Property Owners Association: www.ipo.org

The Licensing Executives Society International: www.lesi.org

United States Copyright Office: www.copyright.gov

United States Patent and Trademark Office: www.uspto.gov

World Intellectual Property Office (Patent Cooperation Treaty): www.wipo.int

BLOGS

www.ipwatchdog.com: news and commentary for all types of intellectual property.

patentlyo.com: news, commentary, statistics, and job listings relating to patents in general.

www.patentbaristas.com: news and commentary relating to biotechnology and pharmaceutical patents.

www.patenthawk.com/blog ("The Patent Prospector"): news and commentary on patents in general.

271patent.blogspot.com ("The 271 Patent Blog"): news and commentary on court decisions in patent cases, plus links to websites for USPTO, Court of Appeals for the Federal Circuit, and the European Patent Office.

www.patentdocs.org: law and news relating to biotechnology and pharmaceutical patents.

www.ipnewsflash.com: news, notices and caselaw from the USPTO, the EPO, the European Trademark Office, the German Patent and Trademark Office, WIPO, and patent offices in Japan, Switzerland, China, as well as courts in Europe, Germany, and the U.S.

www.ipkat.com: copyright, patent, trade mark, information technology, and privacy/confidentiality issues from a UK and European perspective.

blogs.bnip.com/tib ("The Invent Blog"): practical information on patents for small entities.

www.chicagoiplitigation.com: advice to businesses regarding their intellectual property rights.

Patents and Published Patent Applications Cited

Number, title, inventor(s)	Page
2003/0022748, "Device for Closing a Chain, Particularly a Bicycle Chain," Meggiolan	78, 80
2003/0108726, "Chemical Arrays," Schembri et al.	106
2005/0075567, "Ultrasonic Diagnostic Imaging System With Assisted Border Tracing," Skyba et al.	124, 125
2005/0191044, "Backside Rapid Thermal Processing of Patterned Wafers," Aderhold et al.	82, 83
2007/0104027, "Tool for Measuring Perforation Tunnel Depth," Brooks	70, 71
2007/0290070, "Automotive Diesel Exhaust Water Cooled HC Dosing," Hornby	96, 97
2009/0198569, "Method and Apparatus for Presenting Advertisements," Ou et al.	132
3,763,770, "Method for Shearing Spent Nuclear Fuel Bundles," Ehrman et al.	74, 75
3,945,315, "Hydraulic Scrap Shearing Machine," Dahlem et al.	74, 75
4,063,333, "Clothespin," Schweitzer	146–156
4,176,468, "Cockpit Display Simulator for Electronic Countermeasure Training," Marty, Jr.	57
4,203,928, "Process for the Preparation of 2-Nitrobenzaldehyde," Meyer	86
4,225,672, "Method for Producing Maltooligosaccharide Glycosides," Hall	38, 39
4,374,226, "Polycarbonate Having Improved Hydrolytic Stability," Sivaramakrishnan	85
4,491,377, "Mounting Housing for Leadless Chip Carrier," Pfaff	26–29
4,492,779, "Aramid Polymer and Powder Filler Reinforced Elastomeric Composition for Use as a Rocket Motor Insulation," Junior et al.	34
4,527,714, "Pressure Responsive Hopper Level Detector System," Bowman	64, 65
4,683,202, "Process for Amplifying Nucleic Acid Sequences," Mullis	40
4,687,777, "Thiazolidinedione Derivatives, Useful as Antidiabetic Agents," Meguro et al.	103, 104
4,786,505, "New Pharmaceutical Preparation for Oral Use," Lovgren et al.	94, 95
4,853,230, "Pharmaceutical Formulations for Acid Labile Substances for Oral Use," Lovgren et al.	94

First to File: Patents for Today's Scientist and Engineer, First Edition. M. Henry Heines.
© 2014 the American Institute of Chemical Engineers, Inc. Published 2014 by John Wiley & Sons, Inc.

Number, title, inventor(s)	Page
4,879,303, "Pharmaceutically Acceptable Salts," Davison et al.	104, 105
5,026,109, "Segmented Cover System," Merlot	90, 91
5,059,192, "Method of Hair Depilation," Zaias	72, 73
5,178,585, "Chain With Easily Adjustable Number of Links," Lin et al.	81, 82
5,191,573, "Method for Transmitting a Desired Digital Video or Audio Signal," Hair	102
5,601,366, "Method for Temperature Measurement in Rapid Thermal Process Systems," Paranjpe	83, 84
5,694,683, "Hot Forming Process," Teets et al.	61, 62
5,747,282, "17Q-linked Breast and Ovarian Cancer Susceptibility Gene," Skolnick et al.	126, 127
5,813,861, "Talking Phonics Interactive Learning Device," Wood	99, 100
5,822,737, "Financial Transaction System," Ogram	76, 77
5,834,905, "High Intensity Electrodeless Low Pressure Light Source Driven by a Transformer Core Arrangement," Godyak et al.	86
5,872,080, "High Temperature Superconducting Thick Films," Arendt et al.	79
5,899,283, "Drill Bit for Horizontal Directional Drilling of Rock Formations," Cox	143–145
5,950,743, "Method for Horizontal Directional Drilling of Rock Formations," Cox	143–145
5,970,479, "Method and Apparatus Relating to the Formulation and Trading of Risk Management Contracts," Shepherd	127–129
6,113,963, "Treatment of Meat Products," Gurzmann et al.	105
6,321,338, "Network Surveillance," Porras et al.	36, 37
6,355,623, "Method of Treating IBD/Crohn's Disease and Related Conditions Wherein Drug Metabolite Levels in Host Blood Cells Determine Subsequent Dosage," Seidman et al.	123, 124
6,436,135, "Prosthetic Vascular Graft," Goldfarb	46, 47
6,484,203, "Hierarchical Event Monitoring and Analysis," Porras et al.	36, 37
6,553,350, "Method and Apparatus for Pricing Products in Multi-Level Product and Organizational Groups," Carter	131, 132
6,708,212, "Network Surveillance," Porras et al.	36
6,711,615, "Network Surveillance," Porras et al.	36
6,714,983, "Modular, Portable Data Processing Terminal for Use in a Communication Network," Koenck et al.	68–70
6,826,298, "Automated Wafer Defect Inspection System and a Process of Performing Such Inspection," O'Dell et al.	29–31
6,834,603, "Attachment Gusset with Ruffled Corners and System for Automated Manufacture of Same," Price et al.	31–33
6,912,510, "Methods of Exchanging an Obligation," Shepherd	127
6,993,858, "Breathable Footwear Pieces," Seamans	95, 96
7,149,720, "Systems for Exchanging an Obligation," Shepherd	127

(Continued)

Number, title, inventor(s)	Page
7,346,545, "Method and System for Payment of Intellectual Property Royalties by Interposed Sponsor on Behalf of Consumer Over a Telecommunications Network," Jones	129, 130
7,467,680, "Motor Vehicle Hood With Pedestrian Protection," Mason	100, 101
7,521,082, "Coated High Temperature Superconducting Tapes, Articles, and Processes for Forming Same," Selvamanickam	78, 79
7,713,133, "Surface Composition for Clay-Like Athletic Fields," Wolf	92
7,725,375, "Systems and Computer Program Products for Exchanging an Obligation," Shepherd	127
8,071,550, "Methods for Treating Uterine Disorders," Schiffman	101, 102
8,224,669, "Chronic Disease Management System," Peterka et al.	93
8,524,634, "Seed Treatment With Combinations of Pyrethrins/Pyrethroids and Clothianidin," Asrar et al.	91, 92
8,599,898, "Slab Laser With Composite Resonator and Method of Producing High-Energy Laser Radiation," Sukhman et al.	97, 98

Cases Cited

(Continued)

First to File: Patents for Today's Scientist and Engineer, First Edition. M. Henry Heines.
© 2014 the American Institute of Chemical Engineers, Inc. Published 2014 by John Wiley & Sons, Inc.

Index

First to File: Patents for Today's Scientist and Engineer, First Edition. M. Henry Heines.
© 2014 the American Institute of Chemical Engineers, Inc. Published 2014 by John Wiley & Sons, Inc.